Industrial Radiology

NON-DESTRUCTIVE EVALUATION SERIES

Non-destructive evaluation now has a central place in modern technology both as a means of evaluating materials and products as they are manufactured and for confirmation of fitness for purpose while they are in use.

This series provides in-depth coverage of the wide range of techniques that are now available for the non-destructive evaluation of materials. Each volume will contain material that is relevant to final year undergraduates in engineering, materials science and physics in addition to post graduate students, experienced research workers and practising engineers. In some cases they will be written with taught courses in mind, while other texts will be for the qualified engineer or scientist who wishes to become familiar with a new topic at research level.

Series editors

Professor S. Palmer   *Professor W. Lord*
Department of Physics   *Department of Electrical and Computer Engineering*
University of Warwick   *Iowa State University*
Coventry   *Iowa*
UK   *USA*

Titles available:
Numerical Modeling for Electromagnetic Non-Destructive Evaluation
N. Ida

Industrial Radiology
R. Halmshaw

Industrial Radiology

Theory and practice

2nd edition

R. Halmshaw

MBE, Ph.D., ARCS, C. Phys., F. Inst. P., Hon. F. Brit. Inst.
NDT, Hon. F. Indian Soc. NDT

*Consultant; formerly Senior Principal Scientific Officer, Procurement
Executive – Ministry of Defence, Royal Armament Research and
Development Establishment, Sevenoaks, UK*

CHAPMAN & HALL

London · Glasgow · Weinheim · New York · Tokyo · Melbourne · Madras

Published by Chapman & Hall, 2–6 Boundary Row, London SE1 8HN, UK

Chapman & Hall, 2–6 Boundary Row, London SE1 8HN, UK

Blackie Academic & Professional, Wester Cleddens Road, Bishopbriggs, Glasgow G64 2NZ, UK

Chapman & Hall GmbH, Pappelallee 3, 69469 Weinheim, Germany

Chapman & Hall USA, 115 Fifth Avenue, New York, NY 10003, USA

Chapman & Hall Japan, ITP-Japan, Kyowa Building, 3F, 2-2-1 Hirakawacho, Chiyoda-ku, Tokyo 102, Japan

Chapman & Hall Australia, 102 Dodds Street, South Melbourne, Victoria 3205, Australia

Chapman & Hall India, R. Seshadri, 32 Second Main Road, CIT East, Madras 600 035, India

First edition 1982

Second edition 1995

© 1995 R. Halmshaw

Typeset in 10 on 12 pt Palatino by Pure Tech India Ltd., Pondicherry, India

Printed in Great Britain by St Edmundsbury Press, Bury St Edmunds, Suffolk

ISBN 0 412 02780 9

A catalogue record for this book is available from the British Library

Library of Congress Catalog Card Number: 95–69069

∞ Printed on permanent acid-free text paper, manufactured in accordance with ANSI/NISO Z39.48–1992 and ANSI/NISO Z39.48–1984 (Permanence of Paper).

Contents

Contents

Preface to the second edition

Industrial radiography is a well-established non-destructive testing (NDT) method in which the basic principles were established many years ago. However, during 1993–95 the European Standards Organisation (CEN) commenced drafting many new standards on NDT including radiographic methods, and when completed these will replace national standards in all the EC member countries. In some cases these standards vary significantly from those in use in the UK at present. These CEN standards are accepted by majority, not unanimous voting, so they will become mandatory even in countries which vote against them. As most are likely to be legal by the time this second edition is published, they are described in the appropriate places in the text.

The most important new technical development is the greater use of computers in radiology. In the first edition, computerized tomography was only briefly mentioned at the end of Chapter 11, as it was then largely a medical method with only a few equipments having found a place in industrial use. The method depends on a complex computer program and a large data store. Industrial equipments are now being built, although their spread into industry has been slow. Computer data storage is also being used for radiographic data. Small computers can now store all the data produced by scanning a radiographic film with a small light-spot, and various programs can be applied to these data. At present, the method is used largely for storage purposes (archiving) or for producing an enhanced image, but programs are being developed to analyse and interpret the image with the aim of providing a computer print-out of the defects shown in the image.

X-ray television systems (radioscopy), also widely known as 'real-time radiography' (RTR) was covered briefly in the first edition, but it has become increasingly important as television systems, X-ray detectors and computers have developed. For some applications the television image obtainable is now as good as, or better than, the film image.

Some uses have been found for the measurement of Compton back-scatter and there are successful industrial applications which are described in Chapter 11. Microfocus X-ray tubes are now produced commercially with a reliable specification and performance, and are in more widespread use. Some other developments in equipment which appear to be near the marketable stage are briefly described, and the safety aspects of ionizing radiations are updated, based on the latest regulations.

An attempt has been made to examine all the available literature on industrial radiology published since the first edition of the book, particularly the NDT periodicals; new references, where relevant, are incorporated into appropriate chapters.

Ron Halmshaw
1995

Preface to the first edition

Few manufacturing processes are so standardized, automated and rigidly controlled that the product can be guaranteed perfect over large-scale mass production. If structures are to be constructed to meet design requirements and materials are to be used economically and efficiently, some form of testing of the finished product will almost certainly be necessary. Whenever production depends on human skills, human errors creep in and faulty products occasionally occur.

With some small products, samples of production can be extracted and physically tested to destruction without great cost losses; proof tests can be done on a pressure vessel, or vibration testing can be carried out to simulate service conditions, but on many large structures such sampling or proof testing is virtually impossible. Also, if one postulates occasional human errors, sampling will not eliminate the defective items and on many critical components and structures 100% inspection is often desirable. NDT or non-destructive inspection (NDI) are the terms used to describe a wide range of testing techniques designed to produce information about the condition of a specimen without doing any damage to it, i.e. after the testing the fitness of the specimen for use in service is unchanged.

In practice only a few NDT methods directly test a product for fitness for service or directly measure its mechanical or physical properties. Most NDT methods are flaw detection techniques. Thus, one finds a crack at a particular point in a welded structure and determines its length and height, but then it is necessary for someone to decide whether or not that particular crack will prejudice the performance of the structure, i.e. cause premature failure. The term 'non-destructive evaluation' (NDE) is sometimes used for this interpretation stage of the inspection process, but equally it can be argued that NDT ends at the stage where the flaws are quantitatively described, and the final stage is for the designer. NDT is not limited to the finished/assembled stage of production; if materials are to be used to the limits of their strength to satisfy the demands of sophisticated equipment, NDT must be considered for incorporation at every stage of the design–product–service–maintenance cycle. Materials must be examined prior to construction; the design must be such as to enable efficient inspection during manufacture and in-service monitoring should be considered as a possible requirement. In recent years a new term has developed: 'condition monitoring' or 'machinery health monitoring'; this also uses NDT methods.

The well-established 'big five' NDT methods are magnetic particle crack detection, penetrant crack detection, radiography, ultrasonic testing and eddy current testing; there are a dozen or more other methods which have more limited applications to special conditions or materials, and new techniques are constantly being developed. Magnetic and penetrant inspection are essentially surface crack detection methods to increase the efficiency of visual inspection of surfaces; eddy current testing can detect internal flaws in thin material, but only radiography and ultrasonic testing can be applied to the detection of flaws anywhere through the volume of large masses of material. Each has its own advantages and disadvantages, and each requires considerable operator skill to use the method to its best advantage.

NDT is a multidisciplinary subject and as yet there are relatively few professionals adequately trained in the subject; the need for information dissemination, particularly scientifically based data, is especially acute because of the relatively small number of well-trained experts in each particular field.

Industrial radiography has been developing steadily for about 40 years, and as the manufacturers of equipment have broadened the range of their products, a large variety of possible techniques has become available, with a wide range of attainable flaw sensitivities. Generally, good radiographic inspection involves a careful compromise between a number of technique parameters and it is comparatively easy, through lack of knowledge, to use an unsuitable or insensitive technique.

This book has been written to establish the physical basis of industrial radiology, to help users to understand the techniques, to make the correct choice of technique for a particular application easier, and to enable users to get the best results out of radiological inspection. An attempt has been made to provide as much basic data on techniques and performance as possible; although new forms of equipment could possibly require some of these data to be modified slightly, it seems unlikely that major changes in equipment performance will occur which will render present techniques obsolete, if only for physical reasons.

The 'Glossary of Terms used in Radiology' (British Standard BS 2597, 1955) has been used throughout this book as far as possible. According to this glossary, **radiography** is concerned with the production of radio-graphs, that is, the production of an image on photographic film by means of ionizing radiation (X-rays, gamma-rays). Other means of im-aging are also possible: **fluoroscopy** is the name given to the production of a visual image on a fluorescent screen by means of X-rays and by extension, **television-fluoroscopy** is commonly used when a closed-circuit television camera is used to view the fluoroscopic screen image and present it on a conventional television display monitor.

Although neutrons are atomic particles and not ionizing radiation, they penetrate many materials with progressive attenuation and so can be used with suitable conversion screens to produce 'radiographs' ana-logous to X-ray images. **Neutron radiography** is now a well-established term and the appropriate techniques are outlined in Chapter 12.

Ionizing radiation has biological effects and the safety aspects of indus-trial radiology must not be neglected. In this field both medical and industrial radiologists follow the recommendations and terminology in the 'Recommendations of the International Commission for Radiological Units' (ICRU); new units (gray, becquerel, sievert) have officially re-placed the better-known older units (roentgen, curie, rad, rem). As the new units are not yet widely used conversions to the older units have been included.

R. HALMSHAW
1981

Introduction: capabilities and limitations of radiographic inspection

Before starting on the details of the principles and techniques of industrial radiology, it will be useful to offer some general comments on the inherent capabilities and limitations of NDT, particularly in relation to flaw detection. Although most NDT methods are used for many applications other than flaw detection, for example, checking the correctness of construction of assemblies, measuring components and spacings, measuring thickness, coatings etc., flaw detection in both metals and non-metals, particularly in welds, is a major application.

Each NDT method has its own advantages and disadvantages. No method is capable of detecting all the flaws in a specimen and an NDT report which states that, for example, radiographic inspection (or any other NDT method) has shown a specimen to be 'free from all flaws' is both untrue and misleading. The correct statement should be that radiographic (or ultrasonic) inspection 'has not revealed any significant defects'. For every NDT method and technique there is a size of flaw which is unlikely to be detected. Apart from the absolute size of the minimum detectable flaw, there is also a probability factor; large flaws are detected with greater certainty. The most serious defect in an engineering structure, the crack, is a particularly complex defect in that it has length, height, opening width, angle (orientation), shape and position, and all these affect the ability to detect a crack by NDT methods.

Radiographic and ultrasonic testing are the two most widely used NDT methods for flaw detection in structures such as welds, and many attempts have been made to compare their capabilities. Unfortunately, these two methods depend on different basic principles and this makes true comparison difficult. Radiographic flaw detection depends on the difference in transmission of radiation through the flaw and through the background material. In the Figure (part (a)) the transmission along SN through a cavity in the specimen is greater than along SM, and this difference in transmission reaching the film or detector produces an image of the cavity. Similarly, crack detectability depends markedly on the crack opening width, crack height h and crack angle α (the Figure, part (b)). An open crack such as shown in the Figure (part (b) (ii)) is more easily detected than the tighter crack shown in the Figure (part (b) (i)). The quantitative relationships are developed in Chapter 8. Ultrasonic crack detection depends on the reflection and scattering of ultrasonic energy from the crack face (the Figure (part (c)), so the crack opening width does not enter the equation unless the crack is extremely tight and allows some transmission of ultrasonic energy across its faces; this is a

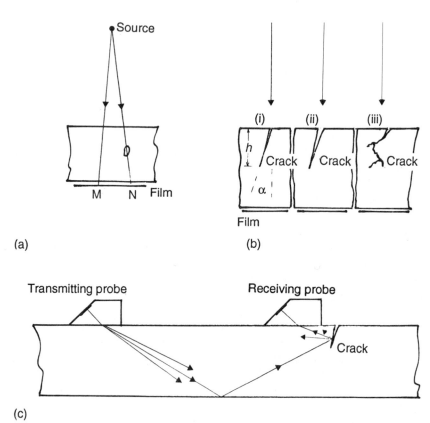

Figure. Crack detection. (a) = basic principle of radiography; (b) = crack detection; (c) = ultrasonic crack detection.

rare condition. Ultrasonic and radiographic crack detection obviously depend on the size (length and height) of the crack.

In both cases, probability and interpretational skills enter the problem. First, a distinction needs to be made between 'was not detected' and 'could not be detected'. All NDT methods, including radiography, are open to occasional wrong interpretation of the image. In addition, with any NDT method, bad techniques with poorer detection capabilities are possible. An ultrasonic beam can be misdirected due to the use of a wrong probe angle or range setting, and serious defects can be missed which would have easily been detected with the correct ultrasonic technique. Radiographs can be taken at the wrong angle to the surface of the specimen, or with a poor technique, for example with too short a source-to-film distance (see Chapter 7). The probability factors in radiography are discussed in Chapter 9.

With a relatively simple flaw shape such as a planar crack, as in the Figure (part (b) (ii)) it is possible, from calculation, to say that a crack of a particular height, width and angle should (or should not) be detected by a specific radiographic technique. With a complex-shaped crack such as in the Figure (part (b) (iii)) such calculations are possible in principle, but have not yet been done. Intuitively, such cracks would be unlikely to be detected by radiography but should have a better chance of detection by ultrasonic testing. In general, with a complex-shaped crack, detection by radiography becomes less and less likely as the complexity of the crack shape increases, whereas with ultrasonic testing, provided there is a receiving probe in the correct location, there will still be some scattered ultrasonic energy to be detected. In principle, therefore, ultrasonic testing should detect cracks which are too small or too complex in shape to be detected by radiography, particularly in thick sections of material. The problems in ultrasonic testing are more those of flaw identification and flaw sizing than flaw detection; the scattered ultrasonic energy from a crack face is very similar to that from any other internal flaw of similar size.

Another complication in making comparisons of ultrasonic and radiographic methods is that as the specimen thickness increases, higher X-ray energies are needed to produce radiographs (see Chapter 7). These higher energies produce greater image unsharpness and so crack detectability becomes poorer. In addition, as the specimen thickness increases, particularly in steel, the cost of suitable radiographic equipment also increases rapidly. With ultrasonic testing, the thickness of the specimen, provided it has good transmission properties and appropriate probe positions are used, does not greatly affect the crack detectability. This applies to most carbon steels, but some austenitic and stainless steels scatter ultrasonic energy strongly and are often anisotropic; these can be extremely difficult to examine successfully with ultrasonics. It also applies to some copper alloys. No such limitations apply to radiographic inspection.

The final points to be made on capabilities and limitations are that radiography-on-film, or radioscopy with a screen image or computer-generated image, produce a 'picture' of the flaws. A flaw is normally easily recognized and interpreted as a crack, slag inclusion, gas cavity etc., and can be measured accurately for length and width, but not for through-thickness height. With ultrasonic testing, length and height are easily measured, but the nature of the flaw is less easily determined, although image analysis of the data by computer programs is developing rapidly. In both radiographic and ultrasonic testing a permanent record of the inspection is easily obtained.

It has been argued in some quarters that as radiography is less satisfactory for the detection of tight or angled cracks, there is no point in using a high-sensitivity radiographic technique and that radiography should be limited to the detection of volumetric flaws; thus, a quicker but relatively low sensitivity technique is sufficient. The Author, in the course of a lifetime of interpreting radiographs, has seen so many cracks discovered by good radiography that he does not accept this philosophy. Also, the cost of the X-ray equipment and the films are the major costs in radiographic inspection whatever technique is used; often the setting-up time is much longer than the exposure time, so the cost and time-saving from using an inferior technique, for the sake of having a very short exposure time, is small.

1

Principles of radiography

Before discussing the properties of ionizing radiation in detail it is desirable to establish the basic principles of radiography. X-rays are generated in an X-ray tube when a beam of electrons is accelerated on to a target by a high voltage and stopped suddenly on striking the target. The X-rays produced have different wavelengths and different penetrating powers according to the accelerating voltage. Gamma-rays have the same physical nature as X-rays and are emitted by certain radioactive substances.

In radiography a source of penetrating radiation (X or gamma) is placed on one side of a specimen and a detector of radiation on the other side. In passing through the specimen the radiation is attenuated as a function of thickness, so that through the thinner parts of the specimen more radiation penetrates and a greater effect is produced at the detector. Therefore, if a near-point source of radiation is used (Fig. 1.1) at a distance from the specimen, a spatial image is produced of the thickness variations in the specimen whether these are due to external thickness changes, internal cavities or inclusions. In most applications the detector is a sheet of photographic film.

Although the essential distinction between radiography and other methods of radiology is in the method of rendering discernible the different intensities of the transmitted radiation, there are certain properties of X-rays which are of common interest. The image on the film is formed by X-rays travelling in straight lines from the source of X-rays, that is, from the focal spot or focus (these terms apply to the electrons, not to the X-rays, which cannot be focused) of the X-ray tube, through the object to the film. This geometric image formation is exactly analogous to shadow formation with visible light, and the sharpness of the image on the film depends on the area of the radiation source and on the relative distances of the object and the source from the film. A small-diameter source, i.e. a small-focus X-ray tube or a small-dimensioned

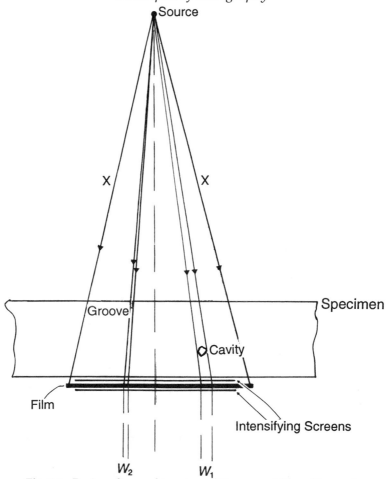

Fig. 1.1 Basic radiographic set-up. ($W_{1,2}$ are widths of images)

gamma-ray source, will be able to produce sharp images. The important dimensions are the effective length and breadth of the focal spot, measured in a plane parallel to the plane of the film. It could be argued that a large-area X-ray focal spot can be compensated for by using a larger distance between the X-ray tube and the film (the source-to-film distance (SFD)). The intensity of the radiation at the plane of the recording film, however, varies inversely as the square of the SFD (the well-known inverse square law), so that as this distance is increased the exposure time, which the film requires to produce a given effect, increases rapidly. If the SFD is doubled, four times the exposure time is necessary, and the exposure time soon becomes too long to be practical.

To produce a satisfactory radiograph it is necessary to match the energy of the radiation to the specimen thickness. Clearly, the penetrating power of the radiation must be such that sufficient radiation is transmitted through the specimen to produce a satisfactory radiograph in a reasonable exposure time. From first principles, however, radiation which is strongly absorbed will produce a greater difference in intensity from a small thickness change in the specimen than will higher energy radiation which is less absorbed. This greater difference in intensity due to a small change in specimen thickness is equivalent to greater radiographic thickness sensitivity, i.e. the ability to detect small changes, so that it is necessary to compromise between sensitivity and penetrating power and not use an unnecessarily high energy of radiation. In practice, this statement on sensitivity and X-ray energy is not completely true because of the complex effects of scattered radiation, but it remains true that the appropriate X-ray energy for a particular specimen thickness must be chosen with regard to both penetration and sensitivity.

There is no absolute relationship between specimen thickness and X-ray energy; it is dependent on the characteristics of the particular X-ray equipment (waveform, output, etc.) and on the sensitivity of the film (and screens) used.

If a film is placed in the X-ray beam behind the specimen for an appropriate time, processed photographically and then placed on an illuminated screen, the differences in X-ray intensity are reproduced as differences in film density (blacknesses), so that different parts of the specimen are seen as different brightnesses. The luminance (brightness) of the illuminated screen must be matched to the density of the radiographic film (Chapter 7).

Instead of recording the variations of intensity of the transmitted X-ray beam on film other ways of recording or displaying this are by:

1. letting the rays fall on to a fluorescent screen of a material which converts X-rays into visible light (this is fluoroscopy);
2. using a device such as an ionization chamber, a Geiger–Müller counter or a scintillation crystal to measure the intensity of radiation at different points behind the object;
3. using an array of certain semiconductor or photodiode materials which are sensitive to ionizing radiation.

There are alternative materials to photographic film, sheets of certain insulating materials for example, which can be used to record an X-ray image. These are described in more detail in Chapter 5.

Factors affecting radiographic image quality

If an image is obtained on a radiographic film the image quality can be assessed in terms of three factors, namely contrast, definition and graininess.

When the film is placed on an illuminated screen for visual inspection, the parts of the film which have received more radiation during exposure, such as the regions under cavities, or thinner parts, appear darker, that is, the film density is higher.

Consider a series of slits, machined in a uniform-thickness specimen. If the difference between the density of the film under a slit, after processing, and the background density is large enough to be discernible, one has an image of the slit. This density difference between the image of the slit and the background is called the image contrast. As a priori statements one can postulate that if the contrast is increased:

1. the image becomes more easily discernible;
2. shallower slits in the specimen will become discernible.

Assuming that the slit has sharp machined edges it could be recorded on the film as either a sharp or a blurred image; this is **image definition** (most radiologists prefer to use the term **image unsharpness**). Intuitively one assumes that a sharper image is of higher quality than an image with a larger unsharpness value; at the limit of discernibility, detection of an image will depend on both contrast and unsharpness.

The third factor in image quality, film graininess, is more complex to explain. When one looks at an image on a radiograph it has a granular appearance and again one assumes intuitively that this granular appearance can mask fine detail in the image.

The three factors of contrast, unsharpness and graininess are the fundamental parameters of radiographic image quality.

One final point which needs to be established at an early stage is the term 'radiographic sensitivity'. This term is universally used by radiographers to describe the ability of a radiograph to show detail in the image, for example, flaws in a metal. If very small flaws can be detected the radiographic image is said to have high (or good) sensitivity. 'Sensitivity' is rarely used to describe the speed of a radiographic process in the way the term is usually used in photography.

2

Basic properties of ionizing radiations

2.1 NATURE OF X-RAYS AND GAMMA-RAYS

X-rays and gamma-rays are forms of electromagnetic radiation having wavelengths roughly in the range 10^{-9}–10^{-13} m. X-rays are produced by allowing a stream of high-energy electrons to strike a metal target, and they originate in the extra-nuclear structure of the target atoms. Gamma-rays are emitted from the nucleus of radioactive elements. Both X- and gamma-rays travel at the speed of light ($c = 3 \times 10^8$ m s^{-1}). They travel in straight lines and are invisible. The essential difference between light, ultra-violet, infra-red, radio-waves and X-rays is one of wavelength and frequency. X-rays and gamma-rays have the property of being able to penetrate matter which is opaque to light and they have a photographic action very similar to that of light. They pass through material of low density more readily than through high density material, and their penetrating ability depends on their wavelength.

The quantum theory, as proposed by Planck, shows that electromagnetic radiation is radiated not continuously but in small 'packets' which he called quanta; it was also necessary to postulate that the photon energy E of each quantum is given by

$$E = h\nu = hc/\lambda \tag{2.1}$$

where h is Planck's constant and ν is the frequency of the radiation. The quantum is, then, the smallest quantity of energy which can be associated with a given phenomenon. It is sometimes more convenient to regard electromagnetic radiation as corpuscular rather than wave-motion and the term photon is used instead of quantum.

2.2 UNITS

The wavelength of X-rays is often quoted in nanometres (1 nm = 10^{-9} m) or in angstroms (1 Å = 10^{-8} cm = 0.1 nm).

The energy of X-rays is also often quoted in terms of the energy of the electrons producing the X-rays (measured in electronvolts (eV)), which is the energy acquired by an electron when accelerated through a potential difference of 1 V. Such units are said to describe the 'quality' of the radiation. In an analogous manner the energy of a gamma-ray is usually given in mega-electronvolts (MeV), although no electron has been used to produce the radiation. In industrial radiography, wavelength is rarely used to characterize radiation; energy or kilovoltage is much more generally used. To understand the properties of X-rays an elementary knowledge of atomic structure is necessary.

2.3 ATOMIC STRUCTURE

An element can be defined as a substance which cannot be broken down chemically into simpler substances. All elements have the same general structure, consisting of a heavy, positively-charged nucleus surrounded by negatively-charged electrons; the electrons are generally thought of as existing in orbits around the nucleus, and their number determines the chemical properties. The nucleus is considered to contain various numbers of protons and neutrons, with the simplest nucleus (hydrogen) consisting of a single proton. As the number of protons in the nucleus increases more electrons can be held in stable orbits around the nucleus and it is these orbital electrons which determine the chemical properties of the individual elements.

Quantum mechanics shows that the orbital electrons are restricted to discrete energy levels and these electron shells are designated by the letters K, L, M, N, ... The K-shell, being closest to the nucleus, contains electrons of the highest energy, and these are consequently the most tightly bound. There are quantum limitations on the number of electrons which may exist in each shell. In the nucleus elements have roughly the same number of protons as neutrons, except for some of the lighter elements, and the number of protons is the atomic number Z. The total number of protons and neutrons (i.e. nucleons) in the nucleus is the mass number, A.

Isotopes are varieties of the same chemical element with the same number of protons but different numbers of neutrons and so different mass numbers. Many of these isotopes are radioactive, or unstable, and are called **radioisotopes**. When an electron is removed from a target atom the atom is left in an unstable 'excited' state and an electron from another electron orbit tends to jump into the vacant space; the difference in energy between the two electron orbits is emitted as a quantum of X-rays. The wavelength of this quantum is characteristic of the target atom and such radiations are called characteristic X-rays. From the nomenclature of the electron orbits these characteristic X-rays are called K-, L-, M- (radiation for the target material) depending on the electron orbit (or orbits) involved in the excitation process. The shortest wavelength K-radiation (uranium K_{β_2}) has a wavelength of 0.11 Å.

2.4 GENERATION OF X-RAYS

In 1895 W. C. Roentgen, Professor of physics at the University of Würzburg, was conducting some experiments on the properties of cathode rays and noticed an effect which had been missed by other investigators. He was working in a darkroom and had his cathode-ray tube in a black box; he noted that a screen coated with barium platinocyanide glowed when brought near the tube.

In his paper [1] describing the new rays, he called them 'X-rays' and noted that they would penetrate a thick book, a 3 mm sheet of aluminium and a thin lead foil. He also mentioned that when he held his hand in front of the screen the outlines of the bones could be seen.

Many laboratories were working with cathode ray tubes (Crookes tubes) and within a month of Roentgen's paper a radiograph of a weld had been published. C.H.F. Muller was already involved in glassblowing and in January 1896 had made the first commercial X-ray tube and produced a radiograph of a human hand. By the end of 1896, the number of applications of X-rays which had already been described was really remarkable.

Roentgen's tube was a high-voltage gaseous discharge tube in which electrons were emitted from a cathode and accelerated towards a target, which they struck with a high velocity. X-rays are emitted whenever matter is bombarded by a stream of electrons.

If an electron starts with zero velocity at the surface of the cathode its kinetic energy on arrival at the target of an X-ray tube, assuming an applied potential difference V between the cathode and target, is

$$\tfrac{1}{2}mv^2 = eV \times 10^7 \qquad (2.2)$$

where m is the mass of the electron, e is the charge on the electron and v is the electron velocity, assumed to be small compared with the velocity of light c.

When an electron with a large kinetic energy strikes the X-ray tube target, there is a transformation of energy. Due to its small mass, the electron is rapidly decelerated in the electric field of the nucleus and the kinetic energy of the electron is partly transformed into a quantum of radiation the minimum wavelength of which is given by

$$\lambda_{min} = \frac{hc}{eV} = \frac{12395}{V} \quad (\text{Å})$$ (2.3 a)

or

$$\lambda_{min} = \frac{1.24}{(kV)} \quad (\text{nm})$$ (2.3 b)

where h is Planck's constant $= 6.624 \times 10^{-28}$ erg s; c is the velocity of light $= 2.998 \times 10^{10}$ cm s^{-1}; e is the charge on the electron $= 4.80 \times 10^{-10}$ esu; m is the mass of the electron $= 9.11 \times 10^{-28}$ g. This radiation is sometimes called 'bremsstrahlung'.

Only a part of the energy of a high-speed electron is required to remove an electron from an atom and the incident electron may lose some energy in this way and then be stopped by direct interaction with the nucleus. The X-ray photon produced has less energy than the original kinetic energy of the electron, i.e. a wavelength greater than λ_{min}. In general, therefore, X-rays of many wavelengths are emitted; the X-ray spectrum is continuous. By analogy with visible light such an X-ray spectrum is sometimes described as 'white' radiation. This continuous spectrum has superimposed upon it the spectrum lines of the characteristic X-rays (Fig. 2.1).

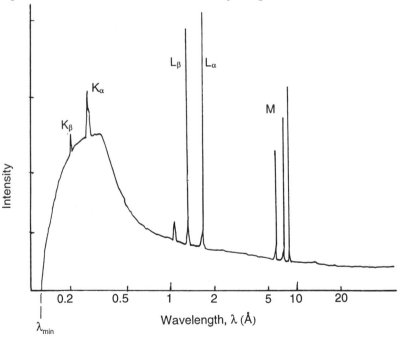

Fig. 2.1 Spectrum from 100 kV X-ray set with no filtration, showing K, L, M characteristic spectrum lines from the target material superimposed on the 'white' spectrum.

By common usage the voltage applied across the X-ray tube is used loosely to describe such a continuous X-ray spectrum. Thus, if the voltage accelerating the electrons is 100 000 V, the X-rays produced would be described as 100 kV X-rays (or sometimes 100 kVp X-rays, the 'p' standing for 'peak'). The use of kilovoltage to describe X-rays is not scientifically accurate as the precise spectrum will depend on the applied voltage spectrum, which is dependent in turn on the type of high-tension generator.

2.5 GAMMA-RAYS

Natural radioactivity was discovered by Becquerel in 1896, and Mme and M. Curie obtained radium chloride from pitchblende ore in 1898. Such natural radioactive substances, in general, emit three types of radiation, originally designated 'α-, β-, γ-rays', and it was later shown that the gamma-rays were electromagnetic radiation of exactly the same physical nature as X-rays, usually of very short wavelength. Artificial radioactive isotopes were first produced by cyclotron bombardment about 1930, and since the advent of atomic piles large numbers of radioisotopes have been identified and studied. The nuclei of these radioisotopes disintegrate with the emission of 'α-, β- or γ-rays', but not necessarily all three forms of emission are produced by any single isotope.

From a radioisotope emitting gamma-rays directly the gamma-ray emission consists of a line spectrum of a series of radiation energies, and the energies and relative intensities are characteristic of the particular radioisotope. The spectrum may be simple or complex. A gamma-ray spectrum consists, therefore, of a series of discrete quantum energies and it is more usual to specify these by energy (keV, MeV) than by wavelength.

It is possible, however, to have a radioisotope source (such as thulium-170) which emits a high energy electron which is reabsorbed in the mass of the thulium pellet, emitting a continuous spectrum of ionizing radiation exactly analogous to the continuous 'white' spectrum from an X-ray tube. The total radiation output of the thulium pellet would be loosely described as 'gamma-radiation'.

2.6 ABSORPTION: ATTENUATION

The intensity of a beam of X-rays or gamma-rays is reduced as the beam passes through matter, i.e. the energy of the radiation beam is attenuated. If this process was one of true absorption, radiography would be a much simpler process, but attenuation includes a multiplicity of processes, some of which result in the production of secondary radiation which may have a different direction of propagation to the primary beam and may also have a different energy. It is usual, therefore, when considering X-rays and gamma-rays, to consider absorption and scattering as two interdependent processes, and it will be more convenient to discuss some of the fundamental processes involved before returning to a discussion of attenuation data and the relationship of scattering processes to practical radiography.

A beam of X-rays or gamma-rays is attenuated exponentially as it passes through a homogeneous thickness of matter, i.e. as the thickness increases the transmitted intensity decreases. For a given incident intensity the emergent intensity decreases, according to an exponential law, with the thickness of the absorber, i.e.

$$I = I_0 \exp\left(-\mu x\right) \tag{2.4}$$

where I_0 is the incident intensity, I is the emergent intensity through thickness x, and μ is a constant which varies in a complex manner with both X-ray energy and the nature of the absorber. Strictly, this equation is true only for the case of narrow-beam geometric conditions and a monoenergetic beam of radiation.

Expressing the attenuation of X-rays in this way, μ is known as the linear attenuation coefficient and the mass attenuation coefficient is μ/ρ where ρ is the physical density of the absorbing material. The term 'cross-section' is also used for various absorption and scattering processes, where the probability that a photon will be affected by the process in traversing a slab of material is called the cross-section for that process. Cross-sections are usually expressed in areas (hence the name) and a common unit is the barn (1 barn = 10^{-24} cm^2). The atomic cross-section is equal to the atomic attenuation (absorption) coefficient.

Scattered radiation

The attenuation of X-rays is not necessarily a simple process in which the primary X-ray energy changes to some form of energy other than ionizing radiation (i.e. true absorption). There may be conversion to X-ray energy of a different wavelength and a different direction of travel, or there may be liberation of secondary atomic particles. The main types of attenuation of concern in industrial radiology are

- photoelectic absorption
- Rayleigh scattering
- Compton scattering
- pair production.

The total attenuation is the sum of the components due to these different effects.

If there are definite phase relationships between the incident radiation and the scattered radiation, the term 'coherent scattering' is sometimes used. If the scattering elements act independently of one another, so that there are no definite phase relationships among the different parts of the scattered beam, the scattering is incoherent. Rayleigh scattering and scattering of neutrons by the regularly spaced atoms of a crystal are coherent: Compton scattering is incoherent.

In photoelectic absorption the photon is absorbed and disappears as a result of the interaction and an atomic electron leaves the atom. The energy of the photon is used in removing an electron from its shell and giving it kinetic energy. The binding energy of the electrons increases with their proximity to the nucleus and with the atomic number of the element (e.g. for lead, 88.6 keV for K-shell electrons). Electrons in the K- and L-shells account for most of the absorption by this process with primary energies greater than the K-edge. Photons with energy very much less than that required to eject an electron are unlikely to be absorbed and consequently the absorption coefficient for the photo-electric effect decreases rapidly as the photon energy increases.

For materials of low atomic number and for photon energies greater than 100 keV, the photoelectric effect has an almost negligible effect in attenuation. For materials of high atomic number the effect is still appreciable even in the megavoltage energy region. The ejected electron flies off approximately sideways in relation to the photon beam for primary photons of energy less than 0.5 MeV, but owing to higher photon momentum, the predominant direction for higher energy photons is more forward.

The atom, which has lost an electron, is said to be 'ionized' and the photoelectron loses energy by producing a series of ion pairs as it travels through the material. The electron shells which lose electrons due to the photoelectric effect fill up with electrons from other shells and the difference in energy level between the two shells results in the emission of characteristic or fluorescent radiation of the atom. This radiation is of comparatively low energy.

Rayleigh scattering

In this process, which is primarily of importance at low photon energies, the incident photon is scattered by the electrons of the atom, without producing any change in the internal energy of the atom or releasing any electrons. The photon is deflected but does not change in energy. Again, the scattering is mainly in the forward direction and becomes more nearly aligned with the primary photon as the energy is increased. The effect is more important for materials of high atomic number.

Compton scattering

The main contribution of scattering to the total attenuation arises from simple Compton effect processes in which the bonds of the atomic electrons can be disregarded. A photon is scattered inelastically and an atomic electron recoils out of the atom. If the photon energy is very high compared with the binding energy of the electron the latter may be regarded as free.

The result of this scattering process is a recoil electron and a secondary photon of reduced energy, travelling in a different direction from both the electron and the incident photon. As the primary photon energy is increased so the probability increases that the recoil electron direction will be close to the direction of the primary photon. The electron never recoils backwards.

For a single scattering event this type of interaction can be accurately computed, but in practical radiography where a photon is passing through a considerable thickness of absorber there will, of course, be multiple Compton scattering, which is much more difficult to compute.

For a single scattering process the Compton relationship is given by

$$\lambda = \lambda_0 + \frac{h}{mc}\,(1 - \cos\theta) \tag{2.5}$$

or

$$\alpha = \alpha_0 \,/\, [1 + \alpha_0\,(1 - \cos\theta)]$$

and

$$\cos\theta = 1 - \frac{2}{(1 + \alpha_0)^2 \tan^2\phi + 1} \tag{2.6}$$

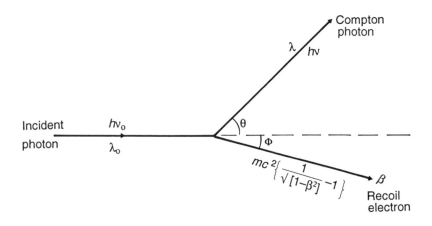

Fig. 2.2 Compton scattering.

where $\alpha_0 = h\nu_0/mc^2$ is the energy of the incident photon in mc^2 ($\lambda_0 = 1/\alpha_0$), and $\alpha = h\nu/mc^2$ is the energy of the scattered photon ($\lambda = 1/\alpha$) (Fig. 2.2). The Compton effect is therefore independent of the atomic number of the scattering material.

The Klein–Nishina law[2] shows that the scattered radiation tends to be more in the forward direction as the incident energy increases. The photon may be deflected through any angle between 0 and 180°, but there is a greater probability that it will travel in the forward direction, and this probability increases as the energy of the primary photon is increased. This forward scatter has the least change in energy from the primary photon. If the photon is deflected through a larger angle the recoil electron has greater energy, but these higher energy electrons have a smaller probability. Over the photon energy range used for industrial radiology, Compton scattering is never negligible. In the lower part of the range it is equally important with photoelectric absorption; at higher energies it is the predominant effect (Fig. 2.3).

Pair production

This effect is only possible with primary photons of energy greater than 1.02 MeV, and the probability of the effect is not appreciable until much higher energies are reached [3]. In this process an individual photon

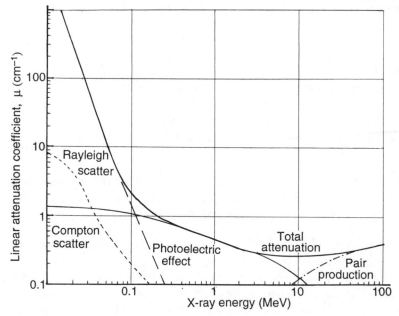

Fig. 2.3 Curves of linear attenuation coefficients for iron plotted against X-ray energy.

disappears, with the creation of an electron–positron pair. Since the mass of the electron is equivalent to 0.51 MeV any excess energy above 1.02 MeV is used to give kinetic energy to the electron and positron. The positrons have an exceedingly short life, and their annihilation is accompanied by the emission of radiation, generally in the form of two photons of energy 0.5 MeV, travelling in opposite directions. The electron may produce bremsstrahlung, so that the attenuation by pair production results in scattered radiation mainly of energy 0.5 MeV.

In measurements of scattered radiation the term 'differential cross-section' is used. It can be defined as the cross-section per unit solid angle for scattering into a small element of a solid angle in a given direction.

2.7 ATTENUATION COEFFICIENT

The general law governing the absorption of radiation as it passes through matter is that the fraction of the radiation absorbed in passing through a thin layer of matter is proportional to the thickness of the layer and to an absorption coefficient, i.e.

$$\frac{I_0 - I}{I_0} = \frac{\Delta I}{I_0} = \mu \, \Delta x \qquad (2.7)$$

and if this is applied to monoenergetic radiation and to an absorber of finite thickness x, it can be integrated to become the exponential relation already given as equation (2.4). As the thickness x is normally measured in centimetres, and as μx must be dimensionless, μ will have dimensions of cm^{-1}. The distance $1/\mu$ is sometimes known as the 'mean free path' of the photon, and for calculations of the depth of penetration the depth is often expressed in relaxation lengths, where $x = 1/\mu$; $\mu x = 1$ is called one relaxation length. Strictly μ should be called the 'total attenuation coefficient' as it represents the attenuation due to the photoelectric effect τ, the Compton effect σ and pair production κ:

$$\mu = \tau + \sigma + \kappa \qquad (2.8)$$

From the definition of μ all contributions to the intensity of the emergent X-ray beam due to scattered radiation should be excluded and the experimental set-up is usually described as a collimated beam, or narrow-beam geometry.

If the attenuation coefficient for a particular material is plotted against the wavelength of the primary radiation, a complex-shaped curve is obtained, which shows characteristic absorption edges at which there is a discontinuous change of absorption. These were first noted by de Broglie in 1916 and they occur when the photon energy of the primary X-rays reaches the binding energy of a particular electron shell. Thus, they can be designated as the K, L, M etc. absorption edges.

It is possible to measure the radiation transmitted through a material and include the scattered radiation emerging from the specimen at directions different from the primary beam. This is contrary to the definition of attenuation coefficient as given in equation (2.8), but nevertheless a value of 'equivalent attenuation coefficient' measured from such data is often quoted. Such an experimental condition is described as 'broad-beam attenuation' or broad-beam geometry (Fig. 2.4(b)), to distinguish it from the condition for measuring the true attenuation coefficient (narrow-beam geometry, Fig. 2.4(a)). It will be obvious that there are many degrees of broad-beam geometry, depending on exactly what proportion of the scattered radiation is included, and that a precise meaning cannot be given, but the condition is of considerable importance in practical radiography as it reproduces the arrangement under which most radiographs are taken. If the specimen is a uniform plate, theoretically of infinite area, and the detector is placed close to the back of the specimen, a film in a cassette for example, then using a large-area beam of radiation, broad-beam conditions are usually taken to mean the experimental arrangement such that an increase in area of the radiation field and the specimen have no effect on the intensity of the transmitted radiation.

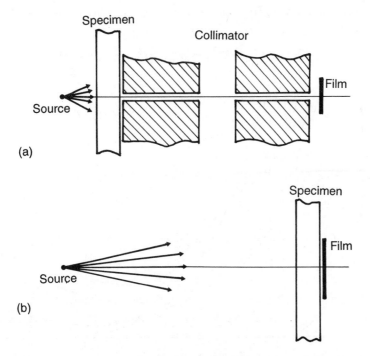

Fig. 2.4 Set-up for measuring the proportion of scattered radiation. (a) = collimated beam, narrow-beam geometry; (b) = open field, broad-beam geometry.

Such measurements are of considerable importance in discussing radiographic techniques and will be considered again later in the book.

Finally, in considering attenuation coefficients it is important to remember that in radiography when 200 kV X-rays is quoted one is not considering a rigidly defined radiation. It is not the same as monoenergetic radiation of energy 0.2 MeV (200 keV); it is a beam of 'white' radiation of maximum energy 200 keV, containing a proportion of all radiation of energy less than 200 keV, the proportions depending on the filtration through which the beam has passed and the characteristics of the X-ray generator and the X-ray tube. The point is of importance here because tables of calculated linear attenuation coefficients are now available and a distinction must be drawn between a calculated value for a monoenergetic beam and a practical X-ray beam as emitted by an X-ray tube. Figure 2.5 shows the calculated values of total attenuation μ for several different materials plotted against energy of radiation.

Half-value thickness

The attenuation can also be expressed in terms of half-value thickness (HVT) or tenth-value thickness, that is, the thickness of material required to reduce the intensity by a factor of two or ten. The following relationships hold:

$$\text{HVT} = T_{1/2} = \frac{0.693}{\mu}, \quad \text{Tenth} \quad \text{VT} = T_{1/10} = \frac{2.303}{\mu} \qquad (2.9)$$

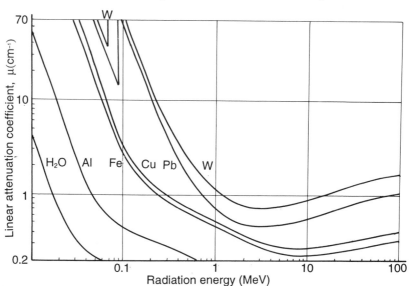

Fig. 2.5 Total linear attenuation coefficients for different materials.

In practice X-ray beams are not monoenergetic, and even if the incident beam was monoenergetic it soon becomes inhomogeneous due to Compton scattering, so that the HVT is not constant for a particular X-ray beam. In defining attenuation coefficient or HVT, the thickness of absorber at which the measurement is made must also be stated.

2.8 ATTENUATION CURVES

A considerable amount of valuable information can be obtained from plots of pairs of experimental attenuation curves for broad- and narrow-beam conditions and for a range of X-ray energies. The experimental set-up is usually as shown in Fig. 2.4 (a) and (b), but Möller and Weeber [4] used an alternative method shown in Fig. 2.6. The two methods give very similar values of the scatter ratio with higher energy X-rays, but at

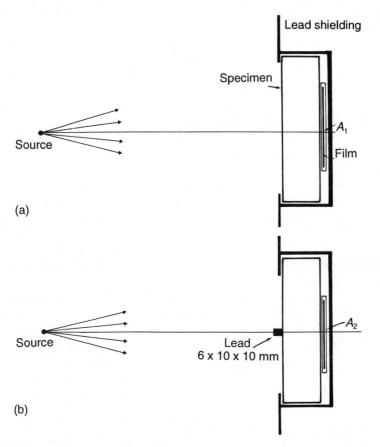

Fig. 2.6 Möller–Weeber method [4] of measuring scattered radiation. (a) = open field, scatter and direct radiation; (b) = scattered radiation only.

low kilovoltages, below 200 kV, the latter method gives lower values, possibly because the blocking bar cuts out some of the scattered radiation as well as the direct beam.

In using the method shown in Fig. 2.4(a) for the narrow-beam conditions, a narrow pencil of radiation is taken but in order to have a measurable area on the film this pencil has to be of a reasonable diameter, usually between 3 and 5 mm, and so may contain some radiation scattered within a small angle of the primary beam direction. The collimating apertures cannot be made very small in practice because of the finite size of the radiation source. In the measurements made by the Author and colleagues, only radiation within an angle of 34 minutes of arc to the primary beam could reach the centre of the film. In experiments with high-energy radiation a greater proportion of Compton scattered radiation is at a small angle to the primary beam and the establishment of narrow-beam conditions becomes more difficult. The elimination of scattered radiation generated in the beam-defining devices needs consideration but can be corrected for if necessary. Similar considerations apply to the broad-beam conditions; as the area of the absorbing plate is increased, the intensity of the scattered radiation measured at the centre of the plate, close to the plate surface, increases towards a limiting value. If the plates are too small, or the measurement is made at some distance away from the plate, a value which is smaller than the true value will be obtained.

Finally, the type of radiation detector to be used requires consideration. The beam of radiation, after absorption and scattering processes, contains a wide range of radiation energies, and detectors with different wavelength-response characteristics may give different results. As the measurements obtained are generally related to experiments or calculations of radiographic sensitivity on film, it seems desirable to employ similar film, with suitable intensifying screens, as the detection medium. This method although laborious and perhaps liable to more experimental error than the use of (say) an ionization chamber, has therefore been used for all the results given in Figs 2.7–2.13.

In Fig. 2.7 the exposure time E for a standard film density is plotted on a logarithmic scale against specimen thickness and this exposure time is taken to be inversely proportional to the intensity of radiation reaching the film. If I_D is taken as the intensity of the direct radiation and I_S the intensity of the additional scattered radiation (i.e. the radiation that is reaching the film under broad-beam conditions and which is travelling in some direction other than along the direct line from the source to the detector) the narrow-beam curve can be taken as a plot of $1/I_D$ against thickness, and the broad-beam curve as a plot of $1/(I_S + I_D)$ against thickness. It can easily be shown that the abscissal separation between the two curves is equal to $\log_{10}[1 + (I_S/I_D)]$. As AC is proportional to $\log_{10}(1/I_D)$ and BC is proportional to $\log_D[1/(I_S + I_D)]$ then

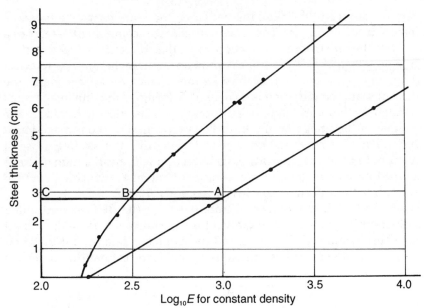

Fig. 2.7 Experimental broad- and narrow-beam curves for caesium-137 radiation and steel plates. AB is a measure of $(I_S + I_D)$.

$$AB = \log_{10}(1/I_D) - \log_{10}[1/(I_S + I_D)]$$

$$= \log_{10}\left(\frac{I_D + I_S}{I_D}\right) = \log_{10}(1 + I_S/I_D) \tag{2.10}$$

This experimental method of determining I_S/I_D will not be directly applicable to salt-screen exposures because of the lack of proportionality between E and I, the reciprocity failure. With salt screens

$$E = \left(\frac{\text{constant}}{I}\right)^x \tag{2.11}$$

where x is likely to have a value of about 0.7–0.8. From this it is easily shown that for salt screens.

$$AB = x \log_{10}[1 + I_S/I_D] \tag{2.12}$$

As salt screens are more sensitive to low-energy radiation one would expect them to intensify the effect of Compton scattered radiation and so increase the values of I_S/I_D compared with lead intensifying screens.

The term $[1 + I_S/I_D]$ is known as the build-up factor (BUF) and is of considerable importance in the calculation of radiographic sensitivity. By definition the direct intensity I_D has been taken to be the component of the radiation intensity reaching the film by travelling on a direct line from the source to the film; this radiation therefore forms the image of defects in the specimen. The scattered radiation component I_S reaches a point on the film from any direction, other than direct; it cannot therefore form an image of a defect at the correct position on the film. As its effect at any one point on the film must be integrated over a large angle, the effect of scattered radiation is an overall 'fogging' of the film image; such radiation is therefore non-image forming and its effect is to reduce the contrast of the image on the film. The proportion of (non-image forming)/(image forming) radiation reaching the film is therefore a useful quantity to be able to control as its reduction will mean a higher image contrast for a given total incident-radiation intensity, which will in turn lead to a better thickness sensitivity (section 2.9).

Considering the fundamental importance of the BUF in controlling radiographic contrast, remarkably little experimental data on the subject have been produced. Practically all the available data are for uniform-thickness plate specimens; attempts to measure scattered radiation on

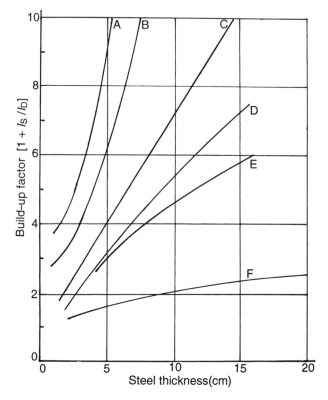

Fig. 2.8 Experimental build-up factor curves for steel plates for various X-ray voltages. A = 200 kV; B = 400 kV; C = 1 MV; D = 2 MV; E = 5 MV; F = 15 MV X-rays.

irregular sections such as pipes have not generally been successful and only limited data are available.

Figure 2.8 shows $(1 + I_S/I_D)$ values for uniform-thickness steel plates for a wide range of X-rays from 200 kV to 15 MV [4, 5, 7]. Figure 2.9 shows similar data for steel plates and gamma-rays from various sources. All these curves were obtained using film and thin lead intensifying screens as the detector.

Additional experimental data for very high energy X-rays (0.5–2 MV) [6] and calculated values are in reasonable agreement. Some discrepancies in experimental BUF values are inevitable as the experimental method is not easy, and also small differences in the X-ray spectrum emitted by different X-ray machines and the screen thicknesses used in the detector can make a considerable difference in the results. The calculation of a BUF value is extremely complex, as one is dealing with an X-ray beam containing a range of X-ray wavelengths and the spectrum changes as it traverses the specimen thickness; in addition, Compton scatter will be

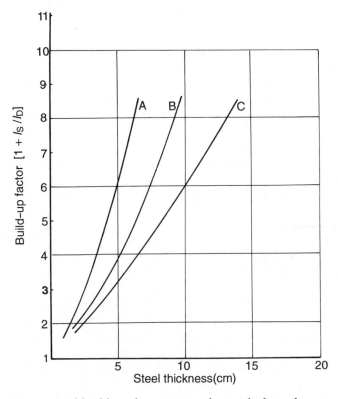

Fig. 2.9 Experimental build-up factor curves for steel plates for gamma-rays from various sources. Curve A = iridium-192; curve B = caesium-137; curve C = cobalt - 60.

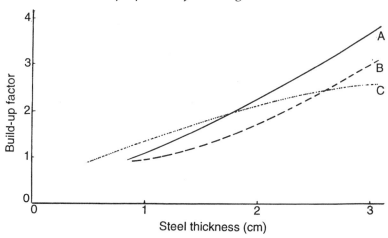

Fig. 2.10 Comparison of build-up factors for steel plates: 200 kV X-rays; no filtration; thin lead intensifying screens. Curves are: Curve A = Halmshaw, experimental [5]; curve B = Möller–Weeber, experimental [4]; curve C = Hirayama, calculated, 2mm film-specimen stand-off [12].

present. Some papers published in Japan [8–12] gave results of numerical computations of scatter for a wide range of geometric conditions, some of which are relevant to conventional radiography. The most rigorous theory leads to a very long computation and was done only for 140 kV X-rays and steel plates. A simplified computation was used for a range of X-ray energies from 50 to 200 keV. Figure 2.10 compares the curve calculated for 200 keV with the Author's and Möller and Weeber's results [4,5], both of which were obtained with film and thin intensifying screens.

Hirayama's data [12] were calculated for a 2 mm film stand-off, which he has shown to make a significant difference to the scatter ratio. There is some discrepancy, but it is unlikely that any experimental technique can determine BUF values to better than ± 20%. One point of interest is that the theoretical curve turns towards a maximum value, which is what one would expect from first principles. The Japanese values provide much new data for low energy X-rays.

From Fig. 2.8 it may be noted that for a given specimen thickness the BUF values are relatively close for 200 and 400 kV X-rays but begin to decrease markedly as one moves to higher energy X-rays, and this decrease in scatter ratio continues to the highest energy (15 MV X-rays) for which measurements are available. This has very important consequences for megavoltage radiography (Chapter 7).

Figure 2.11 shows that the use of thin lead intensifying screens, such as are used for taking radiographs, has the effect of markedly reducing the proportion of scattered radiation and this same effect is obtained with

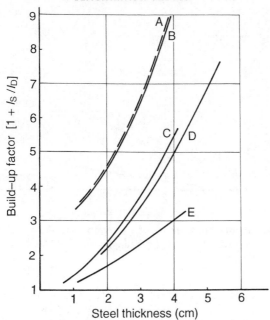

Fig. 2.11 Experimental build-up factor curves for steel plates. Curve A = 200 kV X-rays, no screens; curve B = 400 kV X-rays, no screens; Curve C = 200 kV X-rays, lead intensifying screens (0.05 mm front, 0.1 mm back screen); curve D = 400 kV X-rays, lead screens (0.05 mm front, 0.1 mm back screens); curve E = 200 kV X-rays, lead screens (0.3 mm lead filter close to tube).

the use of a lead filter. It has also been shown that the use of projective magnification, by moving the film away from the specimen, greatly reduces the value of the BUF. Projective magnification of × 4 or greater almost eliminates scattered radiation from the specimen (Fig. 2.12). It will be seen that, except with thin specimens or with projective magnification techniques, the BUF always has a high value, indicating that under almost all radiographic conditions only a small proportion of the radiation reaching the film is direct or image-forming. There are very little experimental data on BUF values in materials other than steel, but there is no reason to suggest that for equivalent absorbing thicknesses the results would not follow the same pattern. It is sometimes suggested that $(1 + I_S/I_D)$ has a linear relationship with thickness, with a constant of proportionality called the 'scattering factor', but the experimental data do not support this simple relationship except within a limited range of radiation energy.

With an absorber of irregular cross-section, or one in which the edge of the specimen is adjacent to the detector, the proportion of scatter reaching the detector may be much greater. Similarly, if there is extraneous scatter from the walls of the laboratory or from other objects in the X-ray

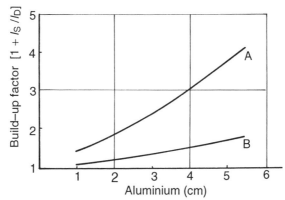

Fig. 2.12 Experimental build-up factor curves for aluminium plates. Curve A = 100 kV X-rays, thin lead intensifying screens; curve B = 100 kV X-rays, projected image, magnification 4:1.

beam, a higher quantity of scatter will reach the film: in these cases any measurements of (I_S/I_D) will be characteristic of that particular set-up and will have no relevance to the general measurement of BUF.

Figure 2.7 is for caesium-137 gamma-rays, which have a monoenergetic spectrum of energy 0.66 MeV, and it can be seen that the narrow-beam 'curve' is a straight line over the whole thickness range investigated. The linear attenuation coefficient calculated from this curve is $\mu = 0.59\,\text{cm}^{-1}$, which corresponds to a radiation energy of 0.62 MeV in steel. The broad-beam curve is continuously changing in slope: it commences with a high slope corresponding to high energy radiation with a low attenuation coefficient, which rapidly changes as scattered radiation is added to the primary beam. For very large specimen thicknesses the slope of the broad-beam curve approaches the slope of the narrow-beam curve.

Figure 2.13 is a corresponding pair of curves for 200 kV X-rays. Here both narrow-beam and broad-beam curves become approximately linear after a longer initial curved portion. In fact, the narrow-beam curve continues to curve gradually upwards as the less penetrating components of the beam are filtered out. It will be observed that in both cases the broad-beam 'curve' is straight, within experimental error, after an initial curved region. This could lead to the conclusion that the beam has become monoenergetic, but this is obviously not the case. In the energy range where Compton scatter is predominant the softer components of the primary beam are constantly being filtered out, but are replaced by multiple Compton scatter produced by absorption of the shorter wavelength components. Thus, an approximate equilibrium is built up which leads to an equilibrium value of the linear attenuation coefficient, which

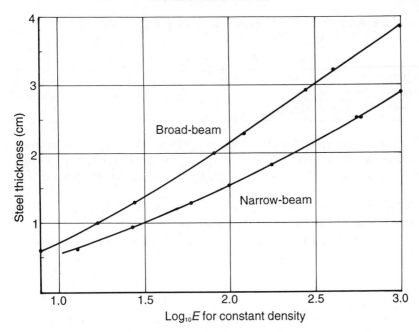

Fig. 2.13 Experimental broad-beam and narrow-beam exposure curves for 200 kV X-rays in steel plates.

can be measured to be almost equal to the linear attenuation coefficient of a monoenergetic beam of energy equal to the maximum energy of the primary 'white' radiation. This has been verified experimentally for kilovoltages above about 1 MV, where the equilibrium value of μ appears to reach about 96% of the quantum limit value.

For lower energy radiation, where photoelectric absorption is predominant, this equilibrium value of μ is approached more gradually as the absorber thickness increases, and is not reached within the normal range of experimental measurements. For X-rays between 20 kV and 250 kV the effective quantum energy (keV), as determined from the slope of the broad-beam exposure curve, is approximately 0.7 times the nominal kilovoltage. At higher energies the quantum kilovoltage approaches more nearly the practical kilovoltage; 400 kV X-rays have an effective quantum energy of about 350 keV, and 300 kV X-rays have an effective energy of 230 keV.

A similar calculation of effective attenuation coefficient for broad-beam geometry with gamma-rays is of considerable interest (Table 2.1). The values in Table 2.1 show the equivalent X-ray energies, in terms of absorption, of various gamma-ray sources when they are used with moderate thicknesses of steel absorbers. They show that in each case the gamma-rays are equivalent to very hard X-rays; for example, although

Table 2.1 Absorption data from gamma-ray sources

Radiation	*Equilibrium μ, broad-beam (cm^{-1})*	*Calculated monoenergetic energy (MeV)*	*Corresponding peak X-ray kilovoltage*
Iridium-192	0.54	0.72	700
Caesium-137	0.435	1.12	1100
Cobalt-60	0.32	2.30	2300

the gamma-ray spectrum of cobalt-60 consists of only two spectrum lines of energies 1.1 and 1.3 MeV, the equivalent X-ray energy, in terms of absorption and therefore of radiographic results, is approximately 2.3 MV X-rays.

Similarly, when used to radiograph moderate thicknesses of steel, iridium-192 gamma-rays are equivalent to 700 kV X-rays. From the viewpoint of practical radiography, the HVT is more immediately useful than the linear attenuation coefficient, and HVTs can be measured on either the broad-beam or narrow-beam curves. As with attenuation coefficients, it is necessary to specify which HVT is being quoted and at what thickness the measurement is made. Sometimes the first HVT is quoted, but a more significant value is the final HVT under broad-beam conditions, i.e. the value when approximate equilibrium has been reached; for steel the maximum value of HVT is about 33 mm. Practical methods of reducing the proportion of scattered radiation are discussed in Chapter 6.

2.9 SCATTERED RADIATION AND RADIOGRAPHIC SENSITIVITY

The effect of BUF and attenuation coefficient in radiography can be shown quantitatively, combined with the photographic factors controlling contrast which are developed in Chapter 5. If the radiographic inspection of a uniform thickness plate specimen is considered it is easily possible to develop an equation for the minimum thickness change in such a specimen which can be detected on a radiographic film.

Let the smallest observable thickness be Δx in a specimen of thickness x; it must be assumed that Δx is small, so that the ratio I_S/I_D is the same for x and $(x + \Delta x)$. On the film the density changes from D to $(D - \Delta D)$ due to the thickness change Δx, and over this small density change the slope of the film characteristic curve (D against log E) can be assumed to be constant, of value G_D, so it is possible to write

$$D = G_D \log_{10} E + K \tag{2.13}$$

where $E = (It)$ is the exposure given to the film, and K is a constant. Then

$$\Delta D = 0.43 \, G_D \frac{\Delta I}{I} \tag{2.14}$$

t being constant. Also,

$$I = I_0 \exp(-\mu x)$$

so

$$\Delta x = -\mu I \, \Delta x \tag{215}$$

Thus, if scattered radiation is neglected

$$\Delta I = -\frac{2.3 \, \Delta D}{\mu G_D} \tag{2.16}$$

If it is supposed that scattered radiation produces a uniform intensity I_S on the film, and that the direct image-forming radiation has an intensity I_D, then it is necessary to put

$$I = (I_S + I_D)$$

into equation (2.14), and $I = I_D$ into equation (2.15), and this gives

$$\Delta x = -\frac{2.3 \, \Delta D}{\mu G_D}\left(1 + \frac{I_S}{I_D}\right) \tag{2.17}$$

If Δx is the minimum thickness change which can be detected, ΔD then must be the 'minimum density difference' discernible by eye on the radiograph, and the ratio $\Delta x / x$ expressed as a percentage becomes the 'thickness sensitivity', also sometimes called 'contrast sensitivity':

$$\frac{\Delta x}{x} \times 100 = \frac{2.3\Delta D\,[1 + (I_S/I_D)]}{\mu G_D x}\, 100\% \tag{2.18}$$

The negative sign has been omitted as it merely means that if Δx is positive, ΔD is negative. It will be noted that μ is the linear attenuation coefficient measured under narrow-beam conditions at the appropriate specimen thickness, not the equivalent attenuation coefficient derived from a broad-beam curve.

Equation (2.18) is of fundamental importance in film radiography as it relates all the parameters of a radiographic technique which control image contrast. It will be seen that the right-hand side of equation (2.18) can be considered to consist of a

- radiation component $= (1 + I_S/I_D)/\mu = K$
- film component $\quad = 1/G_D$
- specimen $\quad\quad\quad = 1/x$
- viewing conditions, i.e. optical component ΔD

These components are discussed at greater length in Chapter 8. The important variable, so far as the radiation energy is concerned, for any fixed value of x is not the BUF alone, but BUF/μ. It can be shown that this ratio rises to a maximum, and begins to decrease for X-ray energies above about 600 keV. At these high X-ray energies, therefore, the attainable thickness sensitivity begins to improve compared with lower energies.

Thickness sensitivity is comparatively simple to measure experimentally with good accuracy, and although the precise value of ΔD to be taken in equation (2.18) is dependent on viewing conditions, it is likely to be constant for one set of conditions and one investigator. As the other parameters in equation (2.18) can be determined reasonably accurately this equation can be used to calculate values of I_S/I_D. This could be particularly valuable at very high X-ray energies in the megavoltage region, where there are little experimental data on BUF values and where the experimental techniques described may not be so accurate because the Compton scatter becomes predominantly small-angle scatter.

Most of this chapter has been concerned with establishing the basic properties of ionizing radiations which are of importance in film radiography. There are many other properties such as, for instance, biological properties, which are very important in health and safety problems, and which are discussed in Chapter 10. The quantum nature of ionizing radiations has been mentioned, but the effects of quantum fluctuations on imaging will be discussed in greater detail in Chapters 8 and 11.

REFERENCES

[1]   Roentgen, W. C. (1898) *Ann. Phys. Leipzig*, **64**, 1.
[2]   Klein, O. and Nishina, Y. (1929) *Z. Phys.*, **52**, 853.
[3]   White, G. (1957) *NBS Circular*, No. 583.
[4]   Möller, H. and Weeber, H. (1963) *Forschungb. Landes Nordrhein-Westf.*, Report 1305.
[5]   Halmshaw, R. (1956) Unpublished Min. of Supply Memo.
[6]   Anon. (1945) *OSRD Report on High Voltage Radiography, No. 4488.*
[7]   Halmshaw, R. (1993) *Brit. J. NDT*, **35**(3), 113.
[8]   Senda, T., Hirayama, K., Yamanaka, H., *et al.* (1987). Study of the scattered/direct radiation intensity ratio on a flat plate specimen in radiographic testing. I. Broad-beam radiation. *Jap. J. NDI*, **36**(6), 411–418 (in Japanese).
[9]   Senda, T., Hirayama, K., Yamanaka, H., *et al.* (1987), Study of the scattered/direct radiation intensity ratio on a flat plate specimen in radiographic testing. 2. Narrow-beam radiation method. *Jap. J. NDI*, **36**(6), 419–425 (in Japanese).
[10] Senda, T., Hirayama, K., Yamanaka, H., *et al.* (1987) Study of the scattered/direct radiation intensity ratio on a flat plate specimen in radiographic testing. 3. Test object with reinforcement. *Jap. J. NDI*, **36**(7), 454–460 (in Japanese).
[11] Senda, T., Hirayama, K., Yamanaka, H., *et al.* (1987) Study of the scattered/direct radiation intensity ratio on a flat plate specimen in radiographic testing. 4. Study of the narrow-beam radiation method in the radiographic testing of welds. *Jap. J. NDI*, **36**(10), 762–771. (in Japanese).
[12] Hirayama, K. (1988) Quantification study on the dose-rate of transmitted radiation and scattered radiation in radiographic testing. Reports 1 and 2. *Trans. Jap. Soc. NDI*, **1**, 1–18.

3

X-ray sources

3.1 INTRODUCTION

X-rays are produced whenever electrons are suddenly brought to rest by colliding with matter. It is necessary therefore, for the following conditions to prevail:

1. a means of producing and sustaining a stream of electrons, i.e. a good vacuum, with a source of electrons which in most modern X-ray tubes is a heated filament;
2. a means of accelerating the electrons to a high velocity, i.e. a means of producing and applying a high potential difference between the source of electrons (the filament) and the target;
3. a 'target' for the electrons to strike, the face of which must also be in the vacuum, aligned with the electron beam, and is therefore an intrinsic part of the X-ray tube.

3.2 X-RAY SPECTRUM

It was shown by Ulrey [1] that if the emission from an X-ray tube is measured and plotted in terms of intensity against wavelength, it has the form of a continuous spectrum with a short-wavelength limit with characteristic spectrum lines superimposed. In industrial radiology, except perhaps specialized low-energy radiography, the characteristic spectrum superimposed on the continuous spectrum is not of great practical importance. If gold is used for the target material, the shortest wavelength characteristic spectrum line (the K_{β_2} line) has a wavelength of 0.15 Å, which is equivalent to an excitation potential of about 80 keV. When a specimen is radiographed, the passage of the radiation through a specimen cuts out the longer wavelengths in the beam by preferential absorption and the spectrum of the radiation reaching the film or detector consists of a small proportion of the extreme short-wavelength end of the primary spectrum from the target, together with scattered radiation generated in the specimen.

Wavelength-intensity distribution in the continuous spectrum

The shape of the continuous spectrum, as plotted in terms of intensity against wavelength, for low energy X-rays of 20–50 kV, is shown in Ulrey's curves (Fig. 3.1). There is a sharp cut-off at the Duane–Hunt limit (λ_0) and the maximum intensity occurs as a broad peak at about $\lambda = 1.5\lambda_0$.

The shape of the continuous spectrum depends greatly on exactly what is measured and plotted. When a scintillation crystal/counter pulse-height analyser system is used, it is usual to plot the number of photons per second per steradian per unit-energy bandwidth (e.g. in successive energy bands 10 keV wide) or photons per square centimetre per kiloelectronvolt per second. Recent investigations of the spectral distribu-

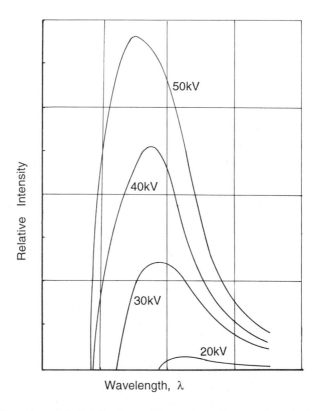

Fig. 3.1 Experimental distribution of intensity with wavelength for different X-ray tube voltages (from Ulrey [1])

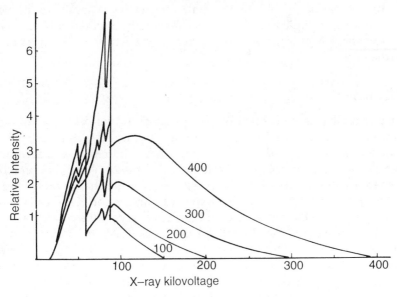

Fig. 3.2 Smoothed experimental intensity–energy distribution curves for 100, 200, 300 and 400 kV X-rays, obtained with a scintillation spectrometer.

tion of the output of X-ray tubes have usually used scintillation spectrometers [2–4] and the type of curve obtained is shown in Fig. 3.2. A collimated beam of very narrow diameter is used to avoid an excessively high counting rate. The channel width of the pulse-height analyser, measured in kilo-electronvolts, will vary with pulse height because of the nonlinear response of the spectrometer to photons, and the pulse-height distribution must be corrected for this effect. The corrected pulse distributions can be transformed into photon spectra by various other corrections which were discussed in detail by Hettinger and Starfeldt [5].

3.3 X-RAY TUBES AND GENERATORS

Having discussed the essential requirements for the generation of X-rays it is therefore convenient to consider the X-ray tube and the generation of the required voltage to accelerate the electrons as separate items.

For the radiations needed for most industrial radiography the accelerating potentials will range from about 30 kV to 30 MV. The electrons acquire the energy equivalent to these voltages from either of two main methods.

1. By direct application of a potential between the source of electrons (or cathode) and the target (or anode). This method has been used in equipment operating up to 2 MV.

2. By application of a relatively small accelerating potential to the electron and arranging for this to be repeatedly applied until the electrons have acquired the desired energy, when they are diverted on to
   the target.

X-ray tubes

Early X-ray tubes developed out of the study of the discharge of electricity through gases when it was found that, as the gas pressure was
reduced, the gas started to conduct and that conduction was accompanied by a glow discharge throughout the tube. When the pressure
reaches a value of about 0.01 mm Hg the glow discharge fades and is
replaced by a brilliant apple-green fluorescence on the glass walls of the
envelope and, as the pressure is still further reduced, the conductance of
the gas decreases until at about 0.001 mm Hg it ceases to conduct and the
fluorescence ceases. This pressure range, 0.01–0.001 mm Hg, is that over
which X-rays are produced and the green fluorescence is due to the
action of the X-rays on the glass envelope. The work of Richardson and
Langmuir on the emission of electrons from hot wires and that of Flem-

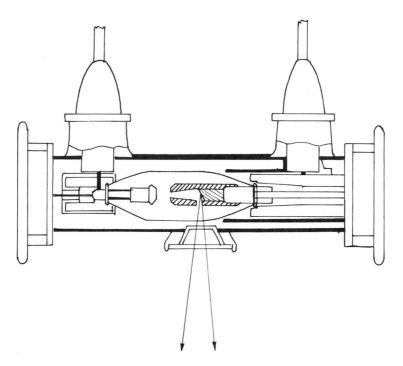

Fig. 3.3 Cross-section of a modern Coolidge-type 300 kV X-ray tube in a
tube-head.

ing on the diode valve led to the Coolidge X-ray tube [6]. In this, the forerunner of modern X-ray tubes, the gas pressure was so low that the residual gas played no part in the production of electrons, a hot wire or filament being the source of supply. A diagram of a Coolidge-type X-ray tube in a modern shielded tube-head is shown in Fig. 3.3. This tube had the great advantages that both the quality and quantity of the emitted X-rays could be kept constant and were independently controllable, the one by the voltage applied across the tube and the other by the current and hence the temperature and electron emission of the filament. The anodes of these early tubes were made of a solid block of tungsten, and since most of the energy of the electrons on striking the target is dissipated in heat, the anode rapidly heated up and in a short time became white-hot. The output of these tubes was therefore limited by the radiant heat cooling of the anode. The demand for still higher outputs and hence the need for more efficient cooling, led to the development of the cooled-anode tube. If the anode is at earth potential this can be water cooling; if not, the cooling system itself has to be insulated.

A medium-voltage X-ray tube usually consists of an insulating, vacuum-tight envelope (generally of glass) containing the cathode and anode. The glass will be of the borosilicate 'hard' glass type, to which it is possible to make glass-to-metal seals using metal alloys of the Kovar type which have the same coefficient of thermal expansion as the glass. The X-ray tube is sealed off under high vacuum and this requires meticulous procedures of pumping and outgasing. Some manufacturers use 'getters' to absorb the last traces of gas. The number of companies making sealed X-ray tubes is probably less than a score, worldwide.

The cathode will consist of a tungsten filament, usually wound spirally, surrounded by a shaped metal electrode; this is the focusing cup. This cup, which is usually made from very pure iron and nickel, acts as an electrostatic lens and controls the shape of the electron beam emitted by the filament. The size of the focal spot depends on the dimensions and location of this cup in relation to the cathode assembly; very accurate assembly is essential. The anode consists of a metal electrode of high thermal conductivity containing the target. Typically, the target is tungsten or gold set in an anode block of vacuum-cast copper. The joint between the copper and the target button must be good, to ensure good heat conduction. The copper anode has a copper extension through the envelope of the X-ray tube and various arrangements are used to dissipate the heat generated.

The face of the target facing the filament is at an angle to the axis of the tube. By this means the energy of the electron beam is dissipated over a considerable area of the target, but seen from the central axis of the X-ray beam, the effective size of the target, the focal spot size, is much smaller

than the real size of the target (Fig. 3.4). This is of fundamental importance in radiographic techniques. By this means, a high tube current, and
therefore a high output of X-rays, can be obtained from a comparatively
small focal spot, without danger of melting the target material. In practice it is common for the electron beam to be focused on to a rectangular
area of target such that the effective focal spot is square. A typical angle
of inclination of the target face is 20°.

X-rays are not emitted with the same intensity in all directions from the
focal spot. Taking the plane of the target as 0°, there is a maximum
intensity at about 32° to this plane (assuming a 20° target) and this
intensity is about 5% higher than that in the central ray. This is known as
the 'heel' effect, but unless the specimen is very large the practical
significance of the effect is small and will be masked by the effects of
variations in the specimen thickness.

X-ray tubes are designed to carry the maximum current possible without melting the target and this will depend on the efficiency of the
cooling system. Usually the anode is hollow and cooling liquid is circu-
lated inside the anode. The whole tube may also be surrounded by
insulating oil and this can also act as a general coolant.

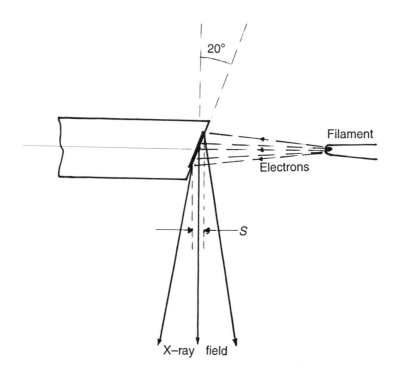

Fig. 3.4 Effective focal spot size (S) of an X-ray tube.

In many X-ray tubes the anode is hooded. This hood consists of a metal body over the anode with an aperture to permit the electron beam to reach the target, and a second aperture for the X-ray beam. The hood also reduces the danger of electron back-emission from the target, if this gets overheated; it reduces the negative charge on the glass wall of the tube, due to secondary electrons scattered from the target and it limits the X-ray emission outside the useful beam, by absorption.

The maximum operating voltage of an X-ray tube depends on the possibility of sparking between either cathode and anode, or surface sparking on the glass. Some tubes use a centre-grounded power supply and the tube is constructed with a metallic central body, electrically grounded; this tube design has proved practical for operation up to about 400 kV but as the applied voltage increases the manufacture of this type of tube becomes increasingly difficult. Several tubes often have to be fabricated to obtain a few satisfactory tubes, and even then premature failure can occur due to charge build-up leading to envelope puncture.

The advent of metal–ceramic tubes with alumina insulators has radically changed this situation [7,8]. The outline design of such a tube, as built by AEG, is shown in Fig. 3.5. The alumina insulators have a flat outer surface and are brazed to each end of a steel cylinder, the tube body; metal tubes fastened at the centre of each insulator carry the cathode and anode assemblies. The most essential parts of the tube are three shielding electrodes, two attached to the metal body and one to the anode. These have to prevent charge build-up by stray electrons on the inner surfaces of the insulators, which might cause instability or even puncture. Secondly, they have to provide a build-up of electric field on the inner surface of the insulator which will accelerate electrons away from these surfaces. In the design of these metal–ceramic tubes the opportunity has also been taken to modify the cable connections (to

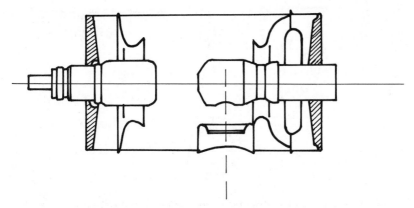

Fig. 3.5 Cross-section of a ceramic X-ray tube.

make them shorter), with the result that a 420 kV tube head can be built so that it is only 70 cm long by 25 cm diameter, weighing about 100 kg. The maximum power quoted is 4.2 kW for a 3.5 × 3.5 mm effective focal spot size. It is claimed that these metal–ceramic inserts are virtually failure-proof. For 420 kV, the tube is operated as a centre-grounded tube but they can be operated as either anode-grounded or cathode-grounded for up to 210 kV. With an anode-grounded configuration, the tube can operate at 20 mA down to 50 kV with a slight increase in focal spot size.

Special types of X-ray tubes

Special tubes with the anode situated at the end of a long pipe-like arm, which are known as 'rod anode' tubes, have been built for the examination of large diameter pipes etc. There are two general types. In one the anode is conical and the electron beam strikes the curved surface of the cone so that a panoramic 360° beam of X-rays is emitted (Fig. 3.6). The electron beam has a considerable path length and is focused by a magnetic lens or a focusing coil; the anode is held at earth potential. The central axis of the X-ray beam is approximately at right-angles to the

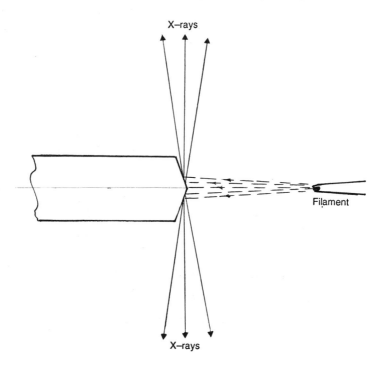

Fig. 3.6 Anode assembly of a rod-anode X-ray tube producing a panoramic beam of X-rays.

tube axis, which is convenient for many applications, such as the inspection of circumferential welds. In the second type of rod-anode tube a flat-faced anode is employed which again gives a panoramic beam of X-rays, but thrown forward of the tube axis.

In a third type of rod-anode tube, a conventional 20° angle face anode is used. In this case the X-ray beam is limited and not panoramic. The development of X-ray crawlers, which carry an X-ray set along a cylindrical pipe, has led to a new interest in rod-anode panoramic-beam X-ray tubes with relatively small focal spot sizes. These crawlers can travel many metres along a pipeline and can be halted at the precise position needed to radiograph a weld. Some crawlers carry in a power line and some operate on batteries. They are made to fit many pipe sizes, from about 150 mm upwards. Crawlers carrying gamma-ray sources are also available.

It is possible to design the filament assembly of an X-ray tube to reduce the effective focal spot size to around 200–300 μm diameter, and these are usually known as minifocus tubes. Some X-ray tubes are dual-focus and can be switched to a minifocus set-up using a separate filament assembly. In the minifocus mode the X-ray output will of course be reduced. Minifocus rod-anode tubes with effective focal spot sizes from 0.4 mm to 2 mm and operating up to 300 kV are available commercially.

Microfocus X-ray tubes

It is possible to use a special electron focusing system between the filament and anode to reduce the effective focal spot size to a few micrometres; these are microfocus tubes. This electron focusing is achieved electrostatically and the microfocus tube is usually of modular construction so that the filament and/or anode can easily be changed; such tubes operate on a turbo-molecular vacuum pump and can be evacuated to the operating condition in about 5 min. Some microfocus tubes are built as sealed tubes. The focal spot size is often adjustable between its minimum size and 200 μm with a matching output for each focal spot size, and microfocus sets have been built to operate from 20 kV to about 400 kV. Commercial microfocus tubes are available with effective focal spot sizes down to 5 μm.

Owing to the dangers of pitting on the target face due to overheating, provision is made either to move the anode to provide a new target area, as required, or deflect the electron beam electromagnetically on to a new area of the anode. Rod-anode microfocus tubes with long anodes up to 3 m long have been built. Due to cooling limitations on the target, the X-ray output of microfocus tubes is necessarily very small and varies with the focal spot size. A typical practical value is 0.1 mA for a 10 μm focal spot at 200 kV.

Microfocus X-ray tubes are designed to have the X-ray source (the target) close to the external surface of the tube, so that a specimen can be placed close to the source; then the film or the detector screen is placed at a distance. A projectively magnified, sharp image is produced. The purpose of these tubes is two-fold:

1. projective enlarged images can be obtained (Chapter 7), without loss of image sharpness;
2. very short source-to-film distances can be used.

The projective method has proved to be particularly useful in radio-scopic systems for overcoming the image definition limitations of fluo-rescent screens (Chapter 11). The use of microfocus and minifocus rod-anode tubes, with a very short source-to-film distance is particularly useful in the inspection of small-diameter welded pipes (Chapter 8).

Multi-section tubes

The chief factor limiting the voltage which can be used in a single-section X-ray tube is the field current effect, and this has led to sectionalized tubes for 1 and 2 MV X-rays. In such tubes the electron beam travels several feet before striking the target and the divergence of the beam is reduced by focusing. The GEC tube for 1 MV X-rays was built as 12 sections of glass tube alternating with Fernico rings. Recently, however, the need for very high energy X-rays in the megavoltage region has led to new forms of X-ray generator, such as linear electron accelerators (linacs), betatrons and microtrons, in which completely different meth-ods are used to generate high-energy electrons and in which conven-tional X-ray tubes are not used.

'Flash' X-ray tubes and generators

An electron beam of very high intensity can be produced by field emission from a small-radius metal cathode (a needle point) by the application of a very high voltage pulse of very short duration. The pulse length must be of the order of a few microseconds and the pulses are generated with a capacitor, a pulse transformer and a high-pressure spark gap, with a very low inductance circuit. This pulse of electrons impinges on a target and generates a very short but intense pulse of X-rays.

Extremely high current densities (10^8 A cm^{-2}) are produced, and if the applied field is increased still further the cathode overheats and vaporizes. Field emission cathodes for flash X-ray tubes usually consist of an array of tungsten needles around a conical target. There are two basic types of flash X-ray equipment.

1. Single pulse tubes, designed to give pulses of the order of 10^{-6} to 10^{-8} s; these are used mostly for ballistic and explosive studies and are available commercially from 100 kV to 1200 kV.
2. Small portable equipments, sometimes battery-operated, designed to give a rapid series of pulses, to build up an appropriate exposure on a static specimen. A typical output performance is 125 mR min^{-1} at 1 m distance. These have originated mostly in Eastern Europe [9, 10]. They can be lightweight and small, but the time taken to build up sufficient exposure for radiography on non-screen film can be quite long. They are sometimes used instead of small mains-operated portable X-ray sets for site work on thin specimens.

The high-voltage pulse generator utilized in many of these systems is a modified Marx-surge generator of coaxial design [11]. Voltage multiplication is accomplished by charging capacitors in parallel and discharging them in series. Spark gaps are used to switch from parallel to series operation. One or two of the gaps are triggered by a low voltage signal and the remainder are then overloaded by the transient voltages within the generator. Capacitors, spark gaps and charging networks are housed in an electrically-grounded steel tank. The steel tank may be pressurized to improve insulation and control the breakdown voltage of the fixed spark gaps, and to produce the fast rise-time needed.

The rise-time of the voltage pulses of a Marx-surge generator is not easily reduced below 15 ns and more complex circuitry such as the Blumlein line generator [12], pulse-charged by a Marx generator, has been used. This consists of three coaxial cylinders which behave like two transmission lines and can produce a 600 kV, 10^5 A, 3 ns pulse.

Ciné-flash systems have been built either with multiple tube assemblies or image intensifiers with a high-speed drum camera, or a rotating prism camera. Frame-rates up to 10^5 frames per second have been claimed [12]. Much larger flash X-ray equipments have been built for special purposes such as ballistic studies, much of it operating in the megavoltage region and sometimes using 'one-shot' X-ray tubes [13].

3.4 X-RAY GENERATOR CIRCUITS

Conventional equipment in the 30–400 kV range originally used step-up high-tension transformers and rectifiers. The X-ray tube may be connected to the transformer or transformers by high-tension cables and bushings, or may be in an insulated tank (gas or oil insulation) which contains all the transformer and rectifier circuitry. Several different types of circuit have been used. In the simplest circuits the X-ray tube itself acts as a rectifier and hence can be used with an alternating high-tension (HT) supply, but this imposes a heavy electrical strain on the tube and it is not recommended for industrial equipment unless other considerations make it desirable. One possible use is for small portable equipment. The classical circuits developed by Kearsley, Graetz, Villard, and Greinacher have now largely been replaced by more modern electronics using multipliers, inverters, thyristrons etc. In many cases, high frequency (~12 kHz) is used. Such circuits are better at maintaining a stable output from a changing input voltage and are physically smaller.

From the viewpoint of radiographic techniques, the most important feature of this type of circuitry is the greatly increased accuracy in kilovolt, milliamp readings and the reproducibility of these, both on one set and with another set of the same nominal circuitry. Nevertheless, the European Standards Organisation (CEN) is preparing a standard (TC-138. Doc. N. 95) specifying methods of measuring the X-ray kilovoltage of commercial X-ray sets. At present, the three measurement methods envisaged are the

1. voltage divider method
2. thick filter method
3. spectrometric method.

The second method is proposed as providing an instrument which any owner of X-ray equipment can use; methods 1 and 3 are to enable X-ray set manufacturers to provide a certificate of performance. For some specialized methods, such as computerized tomography, a highly stable X-ray output is essential.

3.5 HIGH ENERGY X-RAY EQUIPMENT

Tuned-transformer units (resonance transformers)

This type of equipment, based on rather different principles to those previously described, was first developed by Charlton and Westendorp [14], for 1 MV and 2 MV generators. In these units a one-end-earthed air-core transformer in a tank with pressurized Freon-12 or sulphur hexafluoride gas insulation is used to produce the high voltage. To obtain efficient coupling between the windings, both the primary and secondary windings are tuned to the supply frequency. The centre of the stack, normally occupied by an iron core, now accommodates a sectionalized X-ray tube. This type of X-ray set has not been available in recent years due to problems of tube construction, but the machines which were built in the 1940s proved to be extremely reliable, some operating for 30–35 years.

Van de Graaff generator

Van de Graaff generators [8, 15–17] operating on a long sectionalized X-ray tube were first built for industrial radiography around 1943. They operated at 1, 2 and 3 MV. Several sets operated until the early 1990s, but such machines are no longer commercially available.

Electric charges obtained from a DC potential of about 30 kV, supplied by a small transformer and capacitor, are sprayed by a corona discharge from a comb on to a closed moving belt of insulating material. The far end of the belt loop passes over a pulley inside the HT hemisphere at the other end of the generator and gives up its charge, through another comb, to this hemisphere. The voltage is limited only by the insulation and will build up until either breakdown occurs or corona discharge prevents further build-up. If, however, a steady load is imposed by the X-ray tube, an equilibrium condition is set up, and the voltage obtained corresponds to a tube current equal to the rate at which charge is supplied to the belt.

To obtain reasonably high tube currents, high belt-speeds of the order of 1500 m min^{-1} are required. The electron beam from the cathode of the X-ray tube is focused on to the target by means of a magnetic focusing coil, and since the generator gives a true constant potential and hence a uniform electron velocity, a very small focal spot of the order of 0.1 mm is possible.

The betatron

The principle of this machine is to accelerate the electrons in a circular path by using an alternating magnetic field. In a betatron the electrons are accelerated in a toroidal vacuum chamber or 'doughnut', which is placed between the poles of a powerful electromagnet. An alternating current is fed into the energizing coils of the magnet and as the resultant magnetic flux passes through its zero value a short burst of electrons is injected into the tube. As the flux grows the electrons are accelerated and bent into a circular path. The magnetic field both accelerates the electrons and guides them into a suitable orbit and hence, to maintain a constant orbit, these two factors must be balanced so that the guiding field at the orbit grows at an appropriate rate. The acceleration continues as long as the magnetic flux is increasing, that is, until the peak of the wave is reached. At this point the electrons are moved out of orbit, either to the inner or outer circumference of the doughnut, by means of a DC pulse through a set of deflecting coils. The electrons then strike a suitable target. The electrons may make many thousands of orbits in the doughnut before striking the target, so that the path length is very great and the vacuum conditions required are in consequence very stringent.

A further development by Wideroe [18] was the use of magnets with separate poles, so that a DC magnetic flux passes through the orbit space and is cancelled out by appropriate windings in the core or centre-piece. Electrons can then be injected near the negative peaks in the magnetic field-time curve, and the whole flux change between the positive and negative peaks of the flux curve can be used for acceleration. These biased fields give twice the output and also allow the central air gap to be reduced in size. The radiation from betatrons is emitted in a series of short pulses; to increase the mean intensity, some machines operate at higher than mains frequency.

Most betatrons designed for industrial use operate in the energy range 15–31 MeV, but so far as is known there are no longer any such machines operating in the UK. The high megavoltage energy was necessary to obtain a good output, but it had the disadvantage of giving a very narrow field of radiation. Recently, a small transportable betatron, designed in the USSR, has been marketed. This operates at 2, 4 and 6 MeV with a maximum output of 6 rad min^{-1} at 1 m distance. The betatron head weighs only 90 kg, and the focal spot size is said to be 0.2×1.0 mm; the low X-ray output may require the use of fast films or intensifying screens, and the makers recommend the use of fluorometallic screens (Chapter 7) which are said to reduce the exposure by a factor of five.

Linear accelerators (linacs)

An alternative method of producing very high intensity electron beams in the energy range from 1 MeV upwards is the linear electron accelerator [19], usually called the 'linac', and linacs have virtually replaced betatrons as high-energy industrial radiography machines, principally because of their capability to produce very high intensity beams of X-rays. Most linacs operate in the energy range 5–12 MeV, although some have been built for 1–2 MeV and also for as high as 25 MeV.

In the travelling wave linac the acceleration of the electrons to a high energy results from the electrons 'surf-riding' a high-frequency electromagnetic wave travelling in a straight line down an accelerator tube known as the corrugated waveguide (Fig. 3.7). The design is such that within this main waveguide the velocity of the electrons and the phase velocity of the microwaves are equal, so that the electrons stay within the same high frequency phase and continuously receive energy from the microwave field. The first part of the waveguide is sometimes called the 'buncher' section as its function is to induce the electrons to bunch into pulses with a short time-length. Electrons which are injected early in phase are slowed down, and electrons which are late are accelerated slightly with respect to the carrier wave. Today, most accelerators use 10 cm wavelength microwaves (S-band), and for a 5 MeV machine the corrugated waveguide is about 1 m long. A few machines use the 3 cm wavelength (X-band). The high-frequency pulses are generated by a magnetron or klystron coupled to the corrugated waveguide by a system of auxiliary waveguides. Electrons are injected at one end and at the other strike a target, generating X-rays. At the target end of the wave-

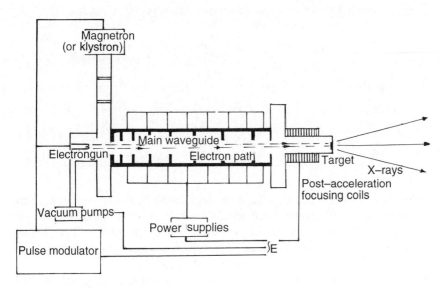

Fig. 3.7 Layout of an electron linear accelerator (linac).

guide it is possible to refocus the electron beam to produce a smaller focal spot and also to increase the X-ray field width.

The output intensity from linacs is many times higher than from beta-trons of the same energy. Thus, from a 12 MeV linac with a 1 mm focal spot, an output of 6000 R min⁻¹ at 1 m is possible, whereas a betatron of the same energy would have an output of only 80 R min⁻¹ at 1m. Most linacs use a vacuum system on the corrugated waveguide, but one manufacturer has used a sealed-off waveguide. In a typical high-output linac the 'head' contains the accelerating waveguide and target, the magnetron and its circuitry and the vacuum pumps. It will weigh about 1200 kg and there are separate modulators, cooling units and controls.

One or two companies have built portable linacs in which the accelerator tube, the microwave unit, the modulator (which supplies the power) and the console are separated. Equipments for 1, 4, 5 and 6 MeV have been built and the radiation head of the largest weighs only 41 kg, much of which is radiation protection. The waveguide connecting the head can be up to 6 m long [20, 21]. X-band microwaves are used instead of S-band. The focal spot is said to be less then 2 mm. Some linacs use the standing-wave system, rather than the travelling wave, in the corrugated waveguide.

Angular intensity distribution of X-rays

With X-ray energies below 1 MeV the practical field size from X-ray tubes is no problem, but with high-energy X-rays the angular intensity

Table 3.1 Field sizes at different X-ray energies, assuming a parallel electron beam

X-ray energy (MeV)	Half-angle of field to 50% intensity (degrees)	Field diameter for 50% intensity (cm)
5	11	39
10	7	24.5
18	5	17.5
31	3.4	11.8
50	1.9	6.7

distribution is of considerable practical importance as the size of the X-ray field can be severely restricted. Several writers have discussed this problem and the formulas proposed by Muirhead and Spicer [22] are generally accepted. The angular intensity distribution controls the size of the X-ray field both from the viewpoint of uniformity and of the maximum useful size of film which can be exposed at any particular source-to-film distance. Thus, if a large source-to-film distance must be used to obtain a useful field size, exposure times will be longer. Practical values of the half-angle to half-intensity are given in Table 3.1 for various energies for which equipment has been built. The experimental values agree closely with calculation. The field diameter to 50% of the central beam intensity, assuming a source-to-film distance of 100 cm, is also included for comparison.

With equipment which works at energies greater than 10 MeV this narrowing of the field makes the field width inconveniently small and it has become common practice to insert a field compensator in the X-ray beam, close to the target. This compensator consists of a piece of metal thicker in the centre than at the edge, so that it absorbs more of the central beam and so equalizes the intensity across the beam; it must be accurately located. The increase in field diameter is, of course, only achieved by a reduction in effective output.

3.6 X-RAY TUBE MOUNTINGS

Except for certain types of routine inspection, for example, longitudinal welds in long lengths of tubing or in welded boiler drums, where a fixed position for the X-ray tube is adequate, it is usually more convenient to be able to move the tube head. This enables radiography of a variety of specimens of different shapes and sizes to be carried out quickly and efficiently. The minimum degree of flexibility that is required is vertical movement and rotation about a horizontal axis, so that the X-ray beam can be directed downwards and then angulated through an angle of at least 90° to give a horizontal beam. In addition, it may also be necessary or convenient to have a certain degree of horizontal movement of the tube head. In this case, a fixed installation on a T-gantry will permit the movements described plus horizontal movement along the gantry arm.

The most essential feature of any mounting is rigidity; the X-ray tube must be stationary during exposures. It is also desirable to have movements in which the position of the X-ray head can be accurately reproduced. An overhead mounting, which does not obstruct the floor in any way, is obviously desirable and motorized controls are also very convenient.

Some equipment is made transportable by mounting the tube stand, the transformers and the controls on a trolley. Equipment with separate HT transformers lends itself more readily to this form of mobile mounting, since the weight of the tube head is so much less than that of the tank unit.

3.7 PORTABLE TANK-TYPE X-RAY SETS

The requirement for radiography of ship's hull welding and welded storage tank inspection led to the development of more truly portable X-ray sets, suitable for on-site inspection conditions. A basic design feature was that the set must be capable of being put through a standard ship's manhole. The first sets were marketed in about 1947 and have been getting steadily smaller and lighter ever since; sets of this type are made for use up to 300 kV. Tank-type X-ray units of this class usually employ self-rectified X-ray tubes, with an earthed cathode. No insulated filament transformer is necessary because the cathode is earthed, and the tank contains the HT transformer and the X-ray tube. Oil or gas insulation is used in the tube/transformer head; if oil is used it can also be pumped around the head to assist in cooling the tube. In most sets there is provision for additional cooling of the oil by an external water supply. Some sets have a limited duty cycle in that they cannot be operated continuously without this auxiliary cooling.

Only a low-voltage connection is required from the tube/transformer tank to the control box and this cable can therefore be made of considerable length without difficulty. An important design point is that both the head and the control box must be waterproof.

3.8 CONTROLS FOR X-RAY EQUIPMENT

Whatever high-voltage circuit is used, provision has if possible, to be made for varying and measuring the high voltage applied to the X-ray tube and the current passing through the tube. The third variable that has to be controlled is the exposure. At a given kilovoltage the integrated output from a given equipment is proportional to the quantity of electricity passed through the tube, that is, the total number of milliampere seconds (mA s).

In some megavoltage machines such as linacs, a radiation-measuring chamber is built into the head of the machine close to the target, which will measure both integrated dose and instantaneous dose rate and convert these to doses at the plane of the film, but such instruments are very rare on lower energy X-ray equipment. A direct method of measuring the total X-ray dose on the specimen or the film is rarely used, but could be of considerable value. It could automatically compensate for slight variations in kilovoltage and tube current during the exposure time.

3.9 COMPARISON OF X-RAY GENERATORS

Table 3.2 gives the physical details of various modern large X-ray sets.

Table 3.2 Comparison of a selection of X-ray sets

Maximum mega voltage	Type	Weight of head (kg)	Focal spot size (mm)	X-ray output (R min^{-1} at 1 m)*
0.250	Tank-type	450	5×5	20 (10 mA)
0.300	Cable-type	91	4×4	40 (8 mA)
0.400	C.P. cables	150	7×6	40 (8 mA)
1	Resonance	1350	7×7	50 (3 mA)
8	Linac	1850	2×2	3000
12	Linac	1850	2×2	4000
6	Minac (portable)	4	2×2	300
24	Linac	2280	2×2	25 000
24	Betatron	3700	0.2×0.2	180
31	Betatron	4500	0.2×0.2	130
6	Betatron	90	0.2×1	3

* Although the roentgen (R) is now a redundant unit and should be replaced by the colulomb per kilogram (= 3876 roentgens) almost all manufacturers of X-ray equipment still quote outputs in R min^{-1} at 1 m

REFERENCES

[1] Ulrey, C. T. (1918) *Phys. Rev.*, **11**, 401.
[2] Heidt, H., Weise H.-P. and Schnitger, D. (1980) *Materialprüfung*, **22**, 211.
[3] Weise, H.-P. (1980) *Materialprüfung*, **22**, 300.
[4] Jost, P., Weise, H.-P., Mundry, E. and Schnitger, D. (1977) *Materialprüfung*, **19**, 426.
[5] Hettinger, G. and Starfeldt, N. (1958) *Acta Radiol.*, **50**, 381.
[6] Coolidge, W. D. (1913) *Phys. Rev.*, **2**, 409.
[7] Reiprich, S. (1979) *Proc. 9th World Conf. NDT*, Melbourne, Australia.

[8] Putzbach, H. (1979) *Proc. 9th World Conf. NDT*, Melbourne, Australia.
[9] Komyak, N. I. and Pelix, E. A. (1979) *Proc. 9th World Conf. NDT*, Melbourne, Australia.
[10] Firstov, V. G., Belkin, N. V. and Lukjanckno, E. A. (1992) *Defektoscopia*, (11), 67.
[11] Mattson, A. (1972) *Proc. 10th Congress High-Speed Photog.*, Nice, France.
[12] Charbonnier, F. (1983) *Non-destructive Testing Handbook*, 2nd edn, section 11, American Society for Non-destructive Testing, Columbus, Oh.
[13] Gilbert, J. F. and Carrick, D. (1974) *Brit. J. NDT*, **16**(3), 65.
[14] Charlton, E. E. and Westendrop, W. F. (1941) *Gen. Elec. Rev.*, **44**, 654.
[15] Van de Graaff, R. J., Compton, K. T. and van Atta, L. C. (1933) *Phys. Rev.*, **43**, 149.
[16] Trump, J. G., Merrill, F. H. and Stafford, F. J. (1938) *Rev. Sci. Instrum.*, **9**, 398.
[17] Trump, J. G., Stafford, F. H. and Van de Graaff, R. J. (1940) *Rev. Sci. Instrum.*, **11**, 54.
[18] Wideroe, R. (1928) *Arch. Electrotech.*, **21**, 387.
[19] Chippendale, T. R. (1961/62) *Philips Tech. Rev.*, **23**(7), 197.
[20] Schoenberg Radiation Corp., Palo Alto, CA (1986) technical literature.
[21] Nakamura, N. and Shiota, M. (1988) *9th Int. Conf. NDE in the Nuclear Industries*, Tokyo, Japan (ed. K. Iida).
[22] Muirhead, E. G. and Spicer, B. M. (1952) *Proc. Phys. Lond.*, **A65**, 59.

4

Gamma-ray sources and equipment

4.1 RADIOACTIVITY

In 1896 Becquerel discovered that certain of the heavier elements emitted penetrating radiation and were unstable; the earlier concept that elements represented the most stable form of matter had therefore to be abandoned. There are three radioactive series known in nature, the parent elements of which are uranium-238, uranium-235 and thorium-232. Each of these decays through a series of daughter elements, which are also radioactive, to a final stable element, which in each series is one of the several isotopes of lead. Radium is one of the daughter elements in the uranium-238 series. The disintegration of the nucleus of a naturally-occurring radioactive substance is accompanied by the emission of one or more forms of radiation which were named alpha-, beta- and gamma-rays. Gamma-rays were shown to be penetrating electromagnetic radiation of the same physical nature as X-rays, and it is the radioactive substances emitting gamma-rays which are used in radiography. Alpha-rays consist of the nuclei of helium and although they may have a considerable kinetic energy they will penetrate only very small thicknesses of material such as, for example, very thin foil. Beta-rays are electrons and also have only a low penetrating power.

It is now well-known that it is possible to have elements with the same atomic number (the number of protons in the nucleus) but different mass numbers (total number of protons and neutrons in the nucleus). Such elements have the same chemical properties and occupy the same position in the periodic table; they are called **isotopes**, and some isotopes have an unstable nucleus and so are radioactive. These are commonly known as radioisotopes or nuclides. Artificially-produced radioisotopes were first introduced in about 1930 by Cockroft and Walton, and the advent of nuclear piles has resulted in production in enormous quantities.

In equations involving radioactive materials it is therefore inadequate to designate an element by its symbol alone, and it is usual to add the mass and the atomic number, i.e.

$$_{27}\text{Co}^{60} \quad \text{or} \quad {}^{60}_{27}\text{Co}$$

although this element is frequently described simply as cobalt-60 (or Co-60).

Disintegration mechanisms

A radioactive atom may distintegrate by one or more of four primary processes which are

1. emission of an alpha-ray (particle)
2. emission of a beta-ray
3. emission of a gamma-ray (or rays)
4. electron capture.

Beta-rays may be positively or negatively charged, i.e. positrons or electrons, and will have energies over a range of values up to a maximum which is characteristic of the particular isotope, i.e. a form of 'white' spectrum of beta-rays. There are some pure beta-emitters, but the majority of beta-emitting nuclides simultaneously emit gamma-rays. A very few radioisotopes decay by simple emission of gamma-rays (a process known as isomeric transition, as there is no change in charge or mass). Gamma-ray emission usually follows promptly after disintegration by one of the other three primary processes, and the gamma-rays range from a few kilo-electronvolts up to several mega-electronvolts in energy (the energy being a characteristic of the radioisotope). Gamma-ray emission may be of a single energy or of several energies, in the form of a number of spectrum lines. Gamma-rays are, of course, of nuclear origin. The nucleus of a radioactive element can capture an orbital electron, and because this electron is usually from the innermost shell, or K-shell, the process is called K-capture. This capture causes a vacancy in the atomic electron structure, which when refilled results in emission of characteristic X-rays.

During disintegration by one of the four primary processes, secondary processes can also occur which cause the emission of additional electromagnetic radiation. If a radioactive nucleus has excess energy, apart from the mechanism described above under isomeric transition, internal conversion can also occur. In this process the energy which would otherwise be emitted as gamma-rays is transferred to an orbital electron, which is ejected. The vacancy is then filled by an external electron resulting in characteristic X-rays, as in the case of electron capture. As an example, in thulium-170 the nucleus first emits a beta-ray, then it emits an 84 keV gamma-ray, or knocks out an orbital electron from the K-shell resulting in 52 keV X-rays. The K-shell internal conversion has a slightly higher probability than the gamma-ray emission. In thulium-170 the beta-ray (electron) has a high energy (900 keV) and is internally absorbed in other thulium atoms, emitting bremsstrahlung radiation by the deceleration of the beta-rays. This radiation has a continuous distribution in energy, i.e. an emergent spectrum of 'white' radiation, with a maximum intensity corresponding to about 40% of the electron energy. This bremsstrahlung radiation forms a major part of the radiation emitted by a thulium-170 source.

4.2 DEFINITIONS

The decay of individual atoms of a radioactive material occurs quite irregularly in time, but the average rate of decay over a period is found to obey statistical laws in which, if a large number of disintegrations are considered, the statistical error will be small.

A general law can be stated. During an interval of time dt (which is small compared with the time required for half the nuclei of a given group to disintegrate), the number which disintegrate dN is proportional to the number of existing nuclei N and to the time dt:

$$dN = -\lambda N \, dt$$

$$N = N_0 \exp(-\lambda t) \tag{4.1}$$

The disintegration constant λ is usually converted to a half-life $T_{1/2}$, the time required for one-half of the nuclei present at a given instant to disintegrate:

$$T_{1/2} = 0.693/\lambda \tag{4.2}$$

Expressed in terms of intensities of radioactivity, i.e. radioactive strengths,

$$I_t = I_0 \exp[(-0.693/T_{1/2})t] \tag{4.3}$$

Thus, after two half-lives the activity is reduced to one-quarter of the original (a 16 Ci source now has the radiation output of a 4 Ci source); after three half-lives the activity is reduced to one-eighth, and so on. The half-lives of different radioisotopes vary from less than a microsecond to many millions of years.

The rate at which a particular radioisotope decays is an important characteristic, and the curie (Ci) is the unit of measure of quantity (strength) most commonly used. Originally, the curie was defined as the disintegration rate of the quantity of radon in equilibrium with 1 g of radium element, but as this definition would be dependent on the accuracy of experimental determinations, it has been redefined internationally as the quantity of any radioactive nuclide in which the number of disintegrations is 3.7×10^{10} s^{-1} [1]. This definition does not take into account the type of radiation or the energy release per disintegration, or the number of particles or photons per disintegration, so that comparison of different radioisotopes in terms of curie-strengths can be misleading.

More recently, the international body responsible for radioactivity units has recommended [2] that the curie is a superfluous quantity and that the new SI unit for the strength of a radioactive source should be the *becquerel (Bq)*, which is equivalent to a reciprocal-second (s^{-1}) so that: 1 Bq = 2.703×10^{-11} Ci, and 1 Ci = 3.7×10^{10} Bq.

As the becquerel is a very small unit, practical radiographic source strengths are given in gigabecquerels (GBq) or terabecquerels (TBq).

A second quantity which is more useful in radiographic work is the radiation output of a given source in terms of roentgens per hour measured at a distance of 1 m from the source (Rhm). A particular radioactive source has a specific Rhm Ci^{-1}. The number of roentgens per hour at 1 cm distance from a 1 mCi source of a gamma-emitter is called the 'K-factor' or the 'specific gamma-ray emission'. The K-factor is therefore ten times larger than the Rhm value. Again, this older unit is much more widely used than the SI unit. The 'specific activity' of a radioisotope source, usually measured in curies per gram, is of importance to radiographers. A higher specific activity means that a source of a given strength can be produced in a smaller physical size, which is of great importance from the viewpoint of radiographic image definition. Also, a small-dimension source has less self-absorption of radiation and so has a greater effective output. The specific activity depends on the neutron flux available for irradiation in the atomic pile and the time the material spends in the pile, as well as on the characteristics of the irradiated material, i.e. the atomic weight and the activation cross-section. Some elements can be activated to a very high specific activity; others are incapable of reaching high activities with presently-available neutron fluxes.

Gamma-ray energy

The usual method of specifying gamma-ray spectrum lines is by their energy in mega-electronvolts rather than by wavelength. The two are related by the quantum relationship

$$eV = h\nu = \frac{hc}{\lambda} \tag{4.4}$$

4.3 PRODUCTION OF GAMMA-RAY SOURCES

Natural sources

Prior to the introduction of artificially-produced radioactive sources radium was used for gamma-radiography. Its great merit was its very long half-life, 1590 years, and its relatively large radiation output. Today it has been completely ousted by the much less expensive artificial isotopes and for industrial use it is only of historical interest.

Radon is a radioactive gas and can therefore be pumped off a radium source, purified and sealed in a capsule. It decays to radium-A, -B, -C, -D etc., and has the same gamma-ray spectrum as a sealed radium source. In the period 1940–55 radon extraction plants were built and radon sources produced in this way [3–6]. A plant containing 4 g of radium in solution can produce a 1600 mC radon source every fourth day. The extracted radon was usually absorbed on a small grain of activated charcoal and could have an effective diameter of less than 0.5 mm. Radon sources have a half-life of 3.8 days and, like radium, are now of historical interest only.

Artificial sources

Although radioactive isotopes can be made in a cyclotron the yield is small and the only methods of production of practical importance at present are neutron activation by irradiation in an atomic pile, and extraction of fission products from the spent fuel elements from an atomic pile. When a neutron is captured by a nucleus a number of different processes are possible, but for the production of gamma-ray sources only the (n, γ) reaction is of interest. This (n, γ) reaction is primarily a thermal neutron reaction and the material produced is a radioactive isotope of the original element, e.g.

$$_{27}Co^{59} + _{0}n^{1} \rightarrow _{27}Co^{60} + \gamma$$

During the activation process, decay of the radioactive isotope is also taking place so that the activity produced tends towards a maximum. Half this maximum is produced in an activation time equal to the half-life, and usually the irradiation time is not longer than two half-lives. As an example, iridium-192 (half-life 74.4 days) irradiated in a neutron flux of 1×10^{11} neutrons cm^{-2} s^{-1}, reaches a specific activity of 175 mC g^{-1} in one week, or 650 mC g^{-1} in four weeks, compared with an activity of 3200 mC g^{-1} at saturation.

Fission products.

When a uranium-235 nucleus captures a neutron, fission generally occurs and a large number of isotopes of different elements are produced, many of which are radioactive. In general, these elements are from the middle range of the periodic table and can be separated chemically from the uranium. This chemical separation has of course to be carried out under conditions such that the radiation hazards are eliminated. A number of gamma-ray sources are obtained in this way in preference to neutron activation methods. For industrial radiography the most important source produced by this method is caesium-137, which occurs in large quantities in used uranium fuel. Several beta-ray emitters are also produced in useful amounts, for example, strontium-90 and krypton-85.

The final material extracted chemically will, of course, be a mixture of all the isotopes which are present and not solely the radioactive isotopes.

4.4 SPECIFIC ISOTOPES FOR RADIOGRAPHY

Although there are several hundred radioisotopes only a few have been widely used for industrial radiography; the remainder are unsuitable for a variety of reasons. The desirable characteristics of a radioisotope for industrial radiography are as follows.

1. The half-life must not be too short; reactivation or frequent exchange of sources is a nuisance and can be costly, because of transport problems.
2. A high specific activity, so as to obtain a compact source with a high gamma-ray output.
3. A high K-factor is also desirable, for the same reason.
4. Low cost of basic material.

The first point is not as important as might be considered at first sight; with good work organization a short-life radioisotope can be used efficiently and economically; witness the use of radon (3.8 days) in the period 1940–55.

It is not easy to generalize on the characteristics of a desirable gamma-ray spectrum. It would be desirable to have a choice of sources each with a spectrum covering part of the energy range from, say, 100 keV to 3 MeV, so as to be able to use a particular source on specimens within a certain thickness range. However, such a choice does not exist at present. The two most frequently used radioisotopes are cobalt-60 and iridium-192. Both emit relatively high energy radiation and would not normally be considered suitable for use on thin sections of steel or for light-alloy radiography.

Cobalt-60 *1.33, 1.17 MeV output*

Cobalt is a hard metal of density 8.9 g cm^{-3} and occurs naturally as a single stable isotope, cobalt-59. By the (n,γ) reaction this is activated to cobalt-60 which decays by emission of a 0.31 MeV beta-particle, followed by two gamma photons in cascade, of energy 1.17 and 1.33 MeV. As the gamma photons are in cascade, they are emitted at the same rates. The half-life is 5.3 years and the radiation output is 1.3 Rhm Ci^{-1}. Cobalt has a high slow-neutron capture cross-section, and due to its neutron absorption characteristics it is usually produced in small pellets or thin discs rather than in a single large source. These pellets or discs are fabricated into sources, after activation, and are also sometimes nickel-plated for protection.

Having a long half-life, cobalt sources need to remain in the atomic pile for relatively long periods to reach high specific activities. Alternatively, special atomic piles with a very high neutron flux are being used for more rapid activation.

Due to their high-energy gammas, cobalt-60 sources require thick shielding for safe handling. A container suitable for a 30 Ci source will be likely to weigh at least 150 kg.

A few cobalt-60 sources of strengths 1000–3000 Ci are in use as static radiation machines; these are used as alternatives to linacs and betatrons for the radiography of large thicknesses of steel from 100–200 mm. They are, of course, very much less expensive than megavoltage X-ray equipment in initial cost, but their radiation output is much smaller so that much longer exposure times are needed than with X-ray machines. Such high-strength sources have considerable self-absorption of radiation: assuming an activity of 120 Ci g^{-1}, a 2000 Ci (nominal) source 10 mm in diameter would be about 24 mm long and the radiation in the forward direction would be equivalent to only about 1360 Ci.

Iridium-192   *0·31, 0·47, 0·6 MeV   (Up to 100 mm ?) Steel.*

Iridium is a very hard metal of the platinum family. Its density is 22.4 g cm^{-3} and it occurs naturally as two isotopes, iridium-191 (38%) and iridium-193 (62%). Activation of natural iridium in the atomic pile is by the (n, γ) reaction and two radioactive isotopes are produced, iridium-192 with a half-life of 74.4 days, and iridium-194 with a half-life of 19 h. It is usual to allow the activated source to decay for several days before measurement, so as to virtually eliminate the short-lived iridium-194.

Iridium-192 decays chiefly by beta-emission to platinum-192 but there is also a second decay process, by electron capture to osmium-192; both these products are stable. The gamma-ray spectrum is complex, containing at least 24 spectrum lines: the principal lines and their relative intensities are shown in Table 4.1 [7–9]. The radiation output is 0.5 Rhm Ci^{-1}.

Table 4.1 Gamma-ray spectrum of iridium-192 (principal lines)

Energy (MeV)	Relative intensity
0.13	0.4
0.28	1.0
0.296	38.0
0.31	37.0
0.32	100.0
0.47	30.0
0.485	1.1
0.59	1.1
0.61	1.9
0.89	0.4

The gamma-ray spectrum is basically in three main groups of energies, around 0.31, 0.47 and 0.61 MeV, with the first group being predominant in relative intensity.

Iridium-192 sources are used for the examination of steel thicknesses up to about 100 mm; for greater thicknesses cobalt-60 will give as good, or better results with shorter exposure times. Iridium-192 is sometimes used for the radiography of steel thicknesses as small as 6 mm, but the results are much inferior in flaw sensitivity to good X-radiographs.

The iridium-192 sources used in industrial radiography are generally between 500 mCi and 50 Ci in strength, but higher strengths up to 100 Ci are available and a 30 Ci source can be as small as 2×2 mm. Containers for Ir-192 sources can be much lighter in weight than those for Co-60 sources, e.g. 12 kg for 40 Ci.

Thulium-170

The need for a radioisotope with a low energy gamma-ray emission, suitable for the radiography of thin steel specimens and for light alloys led to the introduction, in 1953, of thulium-170. Its half-life is 127 days. Thulium is one of the rare-earth elements, and as the metal is difficult to produce it is generally used in the form of thulium oxide; as a powder this can be sintered into pellets of density 7 g cm^{-3}. Thermal neutron capture converts the naturally-occurring isotope thulium-169 into thulium-170. Decay is by emission of 0.968 and 0.884 MeV beta-rays. The excited state of the nucleus is stabilized by emission of 0.084 MeV gamma-ray photons or by internal conversion; only 3.1% of the disintegrations lead to the emission of gamma-rays [10]. From internal conversion a 0.052 MeV gamma-ray is emitted from 5% of the disintegrations. Only about 8% of the total number of disintegrations yield useful low-energy radiation from the viewpoint of radiography. In addition there is a considerable quantity of internal and external bremsstrahlung radiation from absorption of the beta-rays (electrons).

For a 2×2 mm cylindrical source, Liden and Starfelt [11] estimated that the radiation output will consist of 10% 0.084 MeV photons, 45% 0.052 MeV photons and 45% bremsstrahlung photons, the latter having a mean energy of about 0.16 MeV. These proportions depend on the physical size of the source. With a physically smaller source, more electrons will escape and less internal bremsstrahlung will be produced in the source. However, after transmission through any useful thickness of specimen, the bremsstrahlung radiation is likely to be predominant. This has been confirmed by the radiographic results obtained with Tm-170 sources [12]. The bremsstrahlung radiation will tend to become more prominent in the radiation from a larger source; the soft 0.052 and 0.084 MeV photons are more prominent in the emission from very small or very thin Tm-170 sources, so that the actual radiation output depends on the physical size of the source. Output values of 0.0025 Rhm Ci^{-1} have been quoted.

Owing to the preponderance of the bremsstrahlung radiation, thulium-170 sources are not low energy gamma-ray sources as was at first thought, and the radiographic results on thin specimens are not as good as was expected. In addition, it should be noted that very long exposure times are necessary because of the very low radiation output.

Ytterbium-169

This is a more recently available source [13–16]. Ytterbium occurs as a mixture of several stable isotopes, and on neutron activation in an atomic pile Yb-168 activates to Yb-169; the most abundant isotope, Yb-174, activates to Yb-175. The latter has a low energy emission but a half-life of only 4.2 days, whereas Yb-169 has a half-life of 32 days. A newly-activated source, therefore, has a mixture of two radioisotopes with different half-lives and different emission spectra, and the relative proportions of the two will vary rapidly with time. After a relatively short time, most of the Yb-175 will have decayed and the source will be mostly Yb-169. Yb-169 has a gamma-ray spectrum consisting of several spectrum lines, the principal ones as shown in Table 4.2. Some authorities have suggested slightly different relative intensities, and the precise values probably depend on the purity of the isotope.

The radiation output is 125 mR Ci^{-1} h^{-1} at a distance of 1 m. Recently, sources have been made from enriched natural ytterbium, which increases the proportion of Yb-169 in the activated source and so increases the activity which it is possible to obtain in a given physical size. With this enrichment it is possible to obtain about 1 Ci (37 GBq) in a 0.6 mm diameter spherical source and 5 Ci (185 GBq) in a 1 mm source. Smaller

Table 4.2 Ytterbium-169 spectrum

Energy (keV)	Relative intensity
63.4	45
109.9	18
130.6	11
176.9	22
197.7	40
307.4	10

sources are also produced. In the UK the sources are manufactured as spherical ceramic pellets of enriched ytterbium oxide, encapsulated in a miniature source capsule type X.50 (British Standard BS 5288, 1976). Ytterbium is a very rare metal, in short supply, and the enrichment process is difficult, so sources are relatively expensive. The supply situation appears to be the major problem with ytterbium sources, at present.

Experiments measuring the radiographic sensitivity attainable and contrast show that radiographs taken with a Yb-169 source are roughly equivalent to those taken with filtered X-rays between 250 and 350 kV, depending on the specimen thickness [13]. One cannot make a simple statement on the X-ray equivalence of a Yb-169 source; it will depend on the age of the source (because of the mixture of isotopes) and also on the specimen thickness.

Other gamma-ray sources

Many other gamma-ray sources have been proposed for industrial radiography [17] and some have been made available in sufficient strengths for practical assessments of usefulness to be made. The most promising are described below. Beta-emitting isotopes are discussed separately.

Caesium-137 (Cesium in the USA) is a fission product from the atomic pile. It emits a 0.662 MeV single-energy spectrum line and has a half-life of 30 years. The radiation output is 0.37 Rhm Ci^{-1}. It therefore produces radiographs similar to those obtained with Ir-192 and Co-60, but it does not have as high a radiation output intensity and is also more expensive.

Caesium-134 has a half-life of 2.1 years and a complex gamma-ray spectrum of 0.20–1.37 MeV, with the most important lines at 0.6 and 0.8 MeV. Again, the radiographs obtained are similar to those from Ir-192, but the material is not available in high specific activities and it is not widely used.

Xenon-133 is a high-yield gaseous fission product. It is obtained heavily diluted with other gases by the solution of a freshly irradiated uranium slug from the atomic pile. Xe-133, which is a chemically inert gas, is separated from the other gases and can be absorbed on to a pellet of activated charcoal at the temperature of liquid nitrogen. The charcoal pellet is sealed in a metal capsule and it has been shown [20] that a 1 mm Xe-133 source can have an activity of several curies. The half-life is 5.3 days. Xenon-133 has an isomeric form with a half-life of 2.3 days and emits a gamma-ray photon of energy 0.23 MeV in decaying to the ground state. It then decays with a half-life of 5.3 days by the emission of beta-particles (0.35 MeV maximum energy).

The 0.08 MeV gamma-ray is internally converted and 0.03 and 0.035 MeV photons are emitted, leading to the stable form of Cs-133. Of the three spectrum lines, the 0.08 MeV is the most intense. Medical-subject radiographs taken to compare Tm-170 and Xe-133 sources [17–19] show that the latter gives better contrast radiographs.

Europium-152 and 154 is used extensively in Russia, reviews [20] have suggested. It has not been much investigated in western Europe and the parent metal is very expensive. Europium sources usually consist of neutron-activated europium oxide (Eu_2O_3) pressed into pellets. A mixture of three radioactive isotopes is produced; europium-152m, europium-152 and europium-154. Europium-152 has a half-life of 13 years and decays by electron emission to Gd-152 and by K-capture to Sa-152; Eu-154 has a half-life of 16 years and decays by beta emission to Gd-154. The isomeric form has a half-life of only 9.3 h. The gamma-ray spectrum contains a large number of energies and recent investigations give the principal relative intensities as shown in Table 4.3. The specific activity is said to be high and activities of 15 Ci g^{-1} have been produced. Due to the wide range of the gamma-ray spectrum it has been claimed that a europium-152, -154 source will give higher sensitivity radiographs than Ir-192 on steel between 5 and 10 mm thick.

Sodium-24: the target material is usually sodium chloride or carbonate, and activation is by the (n, γ) reaction. Saturation activity is reached in a very short time and the radioisotope has a half-life of only 15 h. The spectrum consists of two very high energy gammas, 1.37 and 2.76 MeV, produced in cascade, and there have been reports of this isotope being used for gamma-radiography of very large thicknesses of material [21]. Applications are only possible where the source can be conveyed very rapidly, from the atomic pile to the site of the radiography.

Selenium-75 is produced from Se-74 by the (n, γ) reaction by irradiation in an atomic pile. The half-life has been variously reported as 118–121 days, and there are several low-energy spectrum lines. The gamma-ray spectrum is stated to be: 115 keV (10%), 137 keV (34%), 265 keV (35%),

Table 4.3 Gamma-ray spectrum of europium-152, -154 source

MeV	0.122	0.244	0.35	0.78	0.87	0.97	1.09	1.28	1.42
Relative intensity	0.2	1.0	2.2	1.3	0.6	1.9	1.9	0.5	2.0

280 keV (15%) and 400 keV (6%). The gamma-ray output is 0.2 Rhm Ci^{-1}. Until 1994 only very low specific activities had been reported, but recently source strengths up to 80 Ci (3 TBq) have been advertised in 3×3.5 mm sources [22], so Se-75 could be a valuable addition to the range of gamma-ray sources for industrial radiography. It is suitable for the examination of thin steel material from about 4 mm up to 35 mm. Special lightweight source containers are proposed, which will weigh about 8 kg for 80 Ci. From the view point of image quality indicators (IQIs) and flaw sensitivities, Se-75 should lie between Yb-169 and Ir-192.

Gadolinium-153 has a half-life of 242 days and principal spectrum line energies of 44 and 100 keV. It has been used for a type of real-time imaging system (Chapter 11).

Iodine-125 is another low-energy source with a number of spectrum lines in the range 0.027–0.035 MeV. The half-life is 57.4 days and very small diameter sources can be produced [23]. It is reported to be very expensive.

Other sources mentioned in the literature at various times are americium-241, europium-155, samarium-153, manganese-54 and cerium-144.

Beta-emitting sources

As a result of the paucity of radioactive isotopes emitting only low-energy gammas and having a reasonable half-life and a high specific activity, several papers [24–27] have drawn attention to the possibility of using bremsstrahlung radiation from beta-emitting isotopes. Possible sources are shown in Table 4.4. The radioactive pellet is sealed in a metal case which is sometimes surrounded by an outer case of Perspex to absorb the beta particles. If an outer case is employed the effective diameter of the source is larger. The sources listed have been shown to

Table 4.4 Beta-emitting isotopes

Source	Energy of beta (MeV)	Half-life (years)
Strontium-90 (parent of yttrium-90)	0.54	28
Yttrium-90	2.26	64.4
Thallium-204	0.77	4.1

be capable of producing good radiographs of thin light-alloy specimens, such as fountain pens and personnel pen-type dosemeters, usually using very short source-to-film distances. The usefulness of these sources depends largely on the activities which can be produced and the availability of the parent materials for activation. Strontium-90 is a fission product; thallium-204 is produced by (n, γ) activation.

4.5 RADIOACTIVE SOURCE-HANDLING EQUIPMENT

A pellet of radioactive material cannot be safely handled except by special methods. Apart from the hazards of the radiation from the source there would be contamination hazards from particles of radioactive matter wiped off by anything touching the pellet. The problems of handling radioactive powder sources are even greater. Artificial radioactive sources for gamma-radiography are always supplied as 'sealed sources', in which the radioactive material is sealed in a capsule by the atomic energy authority of the supplying country, and there should be no need for the industrial radiographer ever to encounter the problems of handling unsealed sources. The sealed radioactive source capsule still cannot, of course, be handled with impunity because of its radiation output. For transportation and handling the radioactive capsule must either be surrounded by a mass of material which absorbs sufficient radiation for the dose-rate on the outside of the container to be reduced to a safe level, or else the capsule must be handled with long tongs, a long rod or by other remote-control handling equipment, so that the distance between source and handler reduces the dose-rate to a safe value. There are three basic methods of providing protection:

1. introducing absorbing material around the source;
2. increasing the distance between the source and personnel (the radiation dose-rate at any point depends inversely on the square of the distance from the source);
3. reducing the time that it is necessary for personnel to be near the source.

Gamma-ray sources are occasionally moved and positioned with the source held on the end of a long rod or string, but this method is only suitable for low-strength sources and is almost obsolete. The usual method is to have the radioactive capsule in some sort of 'exposure container', often called a 'gamma camera', or to use a remotely controlled system such as a crawler to carry a source in a container along the inside of a pipe. The simplest containers essentially have a removable cap or internal mechanism which enables the source to be moved from a 'safe' position (when the radiation dose-rate outside the container is a minimum) to an 'expose' position (in which the full radiation beam is emitted in some specified direction).

There are many designs of such containers. Several systems of source handling are used.

1. For small-strength sources it is possible to design exposure containers which are safe for periods of handling and which are still sufficiently light and small to be portable. The radioactive source may be loaded into such a container by the supplier and it stays in the one container for the whole of its useful life.

2. For larger strength sources, exposing containers have to be used which represent a compromise between safety and weight. Such containers can be handled for only a limited period of time and additional protection must be provided for storage and transport. The safe handling time of such a container may be only a few minutes a day. The exposure is set up with a dummy or empty container and the only handling time of the loaded exposure container is to exchange it for the dummy, a few seconds per exchange.
3. Containers can be designed which are taken close to the exposure site and from which the source capsule is moved to the exposure position, probably into a collimating head, by means of a flexible cable or by pneumatic methods.
4. Very high strength sources are used in large containers which are moved and handled by crane, just like a large X-ray set.
5. For special applications, such as pipeline radiography, a pipeline crawler may carry a gamma-ray source along the inside of a pipe, stop at a predetermined position on the line of a circumferential weld, then open and shut the source container for a chosen exposure time, all by remote control. Much effort has gone into the design of crawlers to go into small pipes. The advantage of this technique is that the whole length of a circumferential weld is radiographed with one film wrapped a round the pipe and one exposure, with the source on the centreline of the pipe.

The accurate location of the crawler along the pipe is a major design problem, the second major difficulty being to keep the mechanism flexible enough to travel round bends and up and down non-horizontal pipework. Sometimes a very small auxiliary gamma-ray source (say Co-57) is carried on the front of the crawler and as this passes under a metal slit fixed on the outside of the pipe with a radiation detector on the line of the slit, there is a sudden increase in the radiation through the slit from this auxiliary source, a pulse of radiation. It is impossible, however, to stop the crawler instantly on receipt of this pulse and allowance must be made for overrun. The design target is to position the source to an accuracy of ± 1 mm; for small-bore pipework even greater accuracies are desirable.

In the UK the dimensions of industrial gamma-ray sources produced by direct irradiation were standardized by the UK Atomic Energy Authority and the British Standards Institution [28]. Some extra capsule designs have been added, to cope with low-energy sources of sub-millimetre sizes. The majority of sources supplied are right-cylinders of diameters equal to lengths of 2, 4 and 6 mm; smaller diameters of 1.0, 0.5 and 0.3 mm are also available and larger strengths are built up from discs. The source is encased in a capsule which has to stand up to rigorous mechanical and physical tests and which also facilitates handling and identification. Other countries such as Canada and USA have their own capsule designs.

A British Standard [29] which is exactly equivalent to an ISO Standard [30] gives constructional requirements for both portable and fixed gamma-ray source containers in terms of exposure-rate limits, safety devices, handling facilities, marking and methods of test. Similar standards and advisory documents are available in other countries, for example, in the USA the NBS Handbook 114: 1975.

The main points to be borne in mind in designing an exposing container are as follows.

1. Provision must be made for loading and exchanging the source capsule easily and quickly.
2. The container must be robust; if accidental damage prevented the container from being properly opened or closed, the handling problem with a large source would become extremely serious. The effect of dirt and moisture on the mechanism must be considered under this heading.
3. A 'fail-safe' design is very desirable, i.e. if anything goes wrong, or is moved during exposure, the source returns to the safe position.
4. There should be a positive indication of the exact position of the source.
5. Except when the container is supported on a mechanized permanent mounting, size and weight are important factors. For a portable container, designed to be put into place by hand, a weight of around 9 kg is about the limit. A design in which there is a lot of waste space around the source capsule, even if this space is occupied by operating mechanisms, is a poor one, as the further the absorbing material is from the source the less protection is afforded by a given weight of container. (If one considers a source inside a hollow lead sphere, the addition of an extra millimetre of lead on the inside surface of the sphere will clearly add less weight than the same extra thickness on the outside.)
6. The containers should be designed so that the beam of radiation can be adjusted in direction; there should also be provision for making panoramic exposures.
7. Scattered radiation, although less of a problem with gamma-rays than with X-rays, should still be kept to a minimum; it is desirable to be able to adjust the beam width to suit the specimen and film size.

There are four basic container designs. The first is shown in Fig. 4.1. The source is at the centre of a sphere of absorbing material of which a conical cap can be removed by a hinged mechanism operated from behind the container. This opening mechanism can also be operated

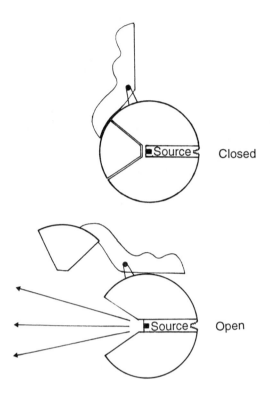

Closed

Open

Fig. 4.1 Gamma-ray source container with hinged cap. The beam width is limited by the cone-angle of the open container.

electrically or mechanically by remote control. The central core of the container, carrying the source, can be extracted or extended for panoramic exposures.

The second design is even simpler (Fig. 4.2). The source, with part of the container, is lifted from the rest of the container and put in the exposing position in an attachment previously fixed in place. The part of the container on which the source is attached provides some protection for the operator. This container design is being phased out.

Figure 4.3 shows an example of a third design, which utilizes a rotating inner cylinder carrying the source. Again, remote operation is possible. This basic design is also often used for static, very high strength sources, such as those using 1000–2000 Ci cobalt-60.

The fourth type of container is quite different and is shown in principle in Fig. 4.4. The source capsule holder is attached to a flexible cable which can be driven along an extension tube. In storage the source is held at the centre of a large storage 'safe' which, if necessary, can be on a wheeled

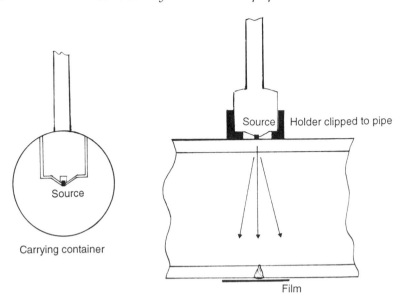

Fig. 4.2 Gamma-ray source container for pipeline work. The source, on its 'handle', is transferred to the exposing position in a holder previously attached to the pipe by a clamp.

vehicle and is at the centre of a U or S-shaped channel so that there is no direct radiation path to the outside. Extension tubing is arranged from the exit port of the storage container to the point where the source is required for making a radiograph, which may be a large distance (10–25 m) from the container. The source is pushed along this extension tube by a winding mechanism which may be hand-operated or electrical; if the latter, it can be coupled to a timing mechanism. The basis of a good design is a source which moves rapidly and smoothly along the flexible extension tube. After exposure the source is moved back into the storage container. Unless required for a panoramic exposure, the source would be moved into a collimating head at the exposure site in order to restrict the radiation beam.

The advantages of this type of container are that a source can be got into awkward places in safety and by remote control; the container can often be kept on the ground. Containers are available for sources up to 100 Ci of cobalt-60. Such sources could not be moved except by remote control; any other pattern of container for such a source would be too large and heavy to be manoeuvred into awkward locations.

One disadvantage of this container design is the vulnerability of the extension tube and cable during the exposure time. If either is damaged it might not be possible to retract the source into its own container.

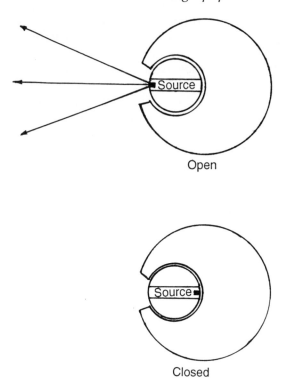

Open

Closed

Fig. 4.3 Gamma-ray source container with rotating core.

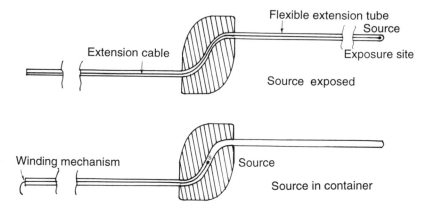

Flexible extension tube

Source

Extension cable

Exposure site

Source exposed

Winding mechanism

Source

Source in container

Fig. 4.4 Flexible-cable type of gamma-ray source container. The bend in the container may be either S- or U-shaped. The winder mechanism is remote from the container on the left-hand side; the source would normally be inserted into a collimating head (not shown) at the exposing position.

With well-designed equipment the time to move a source out to its 'expose' position should be less than 10 s and with a 30 Ci cobalt-60 source, at a distance of 10 m from the source to the control unit, the dose to the operator should be less than 1 mR for each exposure.

For use with ytterbium-169 sources, where one of the most important applications is the inspection of welds in small-diameter steel tubing, special containers have been built both to hold the source on to the outside of the tube for double-wall techniques and to insert the source along the centreline for panoramic exposures, i.e. miniature crawlers. As very small source-to-film distances are used, a physically small source (0.5 mm or less) is necessary, which in turn will also be a low-strength source. This, coupled with the lower energy of ytterbium-169 gamma-rays, means that very small, lightweight containers are possible, in some cases weighing less than 2 kg. The design and use of such small containers for low strength Yb-169 sources is still a new topic. It becomes practical to use two or more sources close together without interference or personnel hazard, and the small size and weight of the containers makes their use in difficult locations much easier than with any radiation source hitherto available.

Until about 25 years ago gamma-ray source containers were usually made of lead cast into a steel or brass shell, but such containers would not meet the present-day thermal test for road transport purposes. Tungsten alloy (tungsten powder in a Monel metal matrix), which has a density of about 16 g cm^{-3} and is machinable, is a more satisfactory material but is now too expensive and many containers are made from depleted uranium in a thin containment shell of steel or brass. The greater the physical density of the material of the container the less the weight for a given degree of protection.

4.6 THE USE OF GAMMA-RAY SOURCES

In general, the choice of a source of radiation for radiography, an X-ray tube or a gamma-ray source, will depend on availability of suitable equipment, thickness of material to be penetrated, accessibility, amount of work to be carried out and the exposure time which can be tolerated. For many applications either X-rays or gamma-rays could produce satisfactory radiographs. A gamma-ray source in a portable container can be used independently of electricity or water supplies at the site of the exposure and can be taken to sites which are difficult to access far more easily than most X-ray sets. However, gamma-rays are not, in general, suitable for the examination of light-alloy materials, although with the wider availability of Yb-169 sources and the advent of Se-75 this may no longer be true. Gamma-ray sources can be used for ferrous specimens up to about 200 mm thick.

Techniques

Many aspects of radiographic techniques with gamma-rays are exactly the same as with X-rays, and these are discussed in detail in later chapters. Some special problems of technique arise with gamma-rays because one cannot control the gamma-ray energy in the way that the kilovoltage of an X-ray tube can be adjusted to match the specimen thickness. The only method of varying the gamma-ray energy is by choosing a different radioisotope.

The two most frequently used radioisotopes, Ir-192 and Co-60, both emit relatively high-energy radiation, so that gamma radiographs usually have a lower contrast than X-radiographs of the same specimens. This difference in contrast becomes less marked as the specimen thickness increases. A good gamma-ray technique attempts to compensate for the lower contrast by ensuring that as high a film contrast as possible is obtained (i.e. higher densities, slower film types).

Gamma-rays are widely used for site radiography, where an X-ray set requiring electrical and water supplies would be awkward or time-consuming to manipulate. For steel specimens less than 50 mm thick, and especially for those less than 25 mm thick, the attainable sensitivity with Ir-192 and Co-60 sources is poorer than on a good X-radiograph, and this loss of sensitivity must be balanced against the convenience of using gamma-rays. For pipe-weld inspection, for example, both X-rays and gamma-rays are employed, and the choice is often governed by considerations of accessibility and working pressures; for high-pressure pipework, X-rays are preferred if their use is possible, but for transmission pipelines, particularly over difficult terrain, gamma-radiography is accepted by many authorities. The CEN: 444: 1994 standard permits the use of Yb-169 gamma-rays on steel down to 1 mm thick for some applications and between 5 and 15 mm for 'more critical' applications, provided fine-grain film is used. A more realistic thickness range would be 2–10 mm of steel. By building a housing which clamps on to a pipe and which can be rotated around the pipe, a series of radiographs of a circumferential weld can be taken rapidly and conveniently (Fig. 4.5) [31]. The source housing acts as a beam collimator as well as personnel protection.

Gamma-rays are extensively used for both ferrous and copper-base casting inspection. The better sensitivity attainable with X-rays is rarely necessary for this type of work, and a medium-strength cobalt-60 source represents a much smaller outlay than a megavoltage X-ray unit for work of thickness beyond the capability of a small X-ray set. Exposure times are, of course, much longer than with X-rays and the volume of work which can be handled may be smaller, but this can sometimes be compensated by careful planning; if numbers of similar thickness

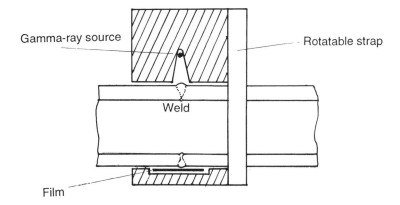

Fig. 4.5 Source holder for the radiography of small-bore pipe welding (the cross-hatched parts are radiation-protective material, e.g. tungsten alloy steel).

castings have to be radiographed these can be oriented around a single gamma-ray source and left for an overnight exposure. By using several tiers of castings, 100 or more radiographs may be exposed simultaneously in one overnight exposure; if these are set up in a locked room there need be no radiation hazard and no supervision. Consequently, most of the day shift is spent on film processing, interpretation of the films and setting up the next series of castings. If many exposures are required on one complex casting, gamma-radiography may be at a considerable disadvantage compared with X-rays in terms of the time required to make a complete examination.

It is also possible to have a rotating turntable carrying a ring of specimens (Fig. 4.6); the required exposure time is adjusted to be one rotation of the table, and as the specimens pass point A, they and the film cassettes are changed. A masking block B enables this to be done in safety and the whole assembly is in a lead-lined room.

A gamma-ray source represents a much smaller capital outlay than a corresponding X-ray set for the radiographic department which does not require maximum flaw sensitivity and which does not need to produce large numbers of radiographs. The attainable flaw sensitivity on thin materials will not generally be as good as can be obtained with X-rays, except perhaps when using a Yb-169 source, but for steel thicknesses above 10 mm the convenience often outweighs the loss of some flaw sensitivity.

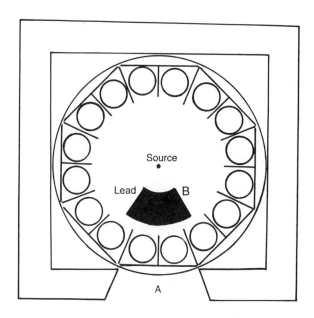

Fig. 4.6 Semi-automatic gamma-ray exposure set-up for cylindrical specimens. The time of rotation of the turntable carrying the specimens is arranged so that the specimens are removed and replaced together with the film cassette as they reach point A, which is shielded from the source by a large block of lead.

REFERENCES

[1]   International Commission on Radiological Units (1950) Recommendations, London.
[2]   International Commission on Radiological Units, (1975) *Am. J. Roentgend.* **125**(2), 492.
[3]   Oddie, T. H. (1937) *Brit. J. Radiol.*, **10**, 348.
[4]   Dawson, J. A. T. (1946) *J. Sci. Instrum.*, **23**, 138.
[5]   Russ, S. and Jennings, W. A. (1948) *Radon: Its Technique and Use*, Murray, London.
[6]   Rose, J. E. and Swain, R. W. (1950) *J. Nat. Cancer Inst.*, **10**, 605.
[7]   Muller, D. E. (1952) *Phys. Rev.*, **88**, 775.
[8]   Kelly, W. A. and Wiendenbeck, M. L. (1956) *Phys. Rev.*, **102**, 1130.
[9]   Halmshaw, R. (1954) *Brit. J. Appl. Phys.*, **5**, 238.
[10] Graham, R. L. (1955) *Can. J. Phys.*, **30**, 459.
[11] Liden, K. and Starfelt, N. (1955) *Brit. J. Appl. Phys.*, **6**, 252.
[12] Halmshaw, R. (1955) *Brit. J. Appl. Phys.*, **6**, 8.
[13] Halmshaw, R. and Manly, T. J. (1971) *Brit. J. NDT*, **13**(4), 102.
[14] Dobrowlski, M. and Jedrzejewski, A. (1975) *Brit. J. NDT*, **17**, 15.
[15] Thiele, H. and Proegler, H. (1976) *Proc. 8th World Conf. NDT*, Cannes, France, paper 3D12.

[16] Pullen, D. and Hayward, P. (1979) *Brit. J. NDT*, **21**(4), 179.

[17] West, R. (1953) *Nucleonics*, **11**(2), 20.

[18] Wilson, E. J. and Dibbs, H. P. (1958) *Nucleonics*, **16**(4), 110.

[19] Mayneord, W. V. and Ireland, H. D. (1956) *Brit. J. Radiol.*, **29**, 277.

[20] Goryacheva, K. G. (1958) *Nauch. Dokl, Vyssh. Scholy. Machino i Pribo.*, (3), 163.

[21] Hinsley, J. F. (1959) *Non-destructive Testing*, p. 405. Macdonald & Evans, London.

[22] Isotopen-Technik Dr Sauerwein GmbH, Haan, Germany (1994) Technical literature.

[23] Griverman, J. and Schivlev, J. (1963) *Nucleonics*, **21**(5), 96.

[24] Green, F. L. and Check, W. D. (1960) *Non-destructive Testing*, **18**(6), 382.

[25] Keriakes, J. G. Krebs, A. T. (1955) *Non-destructive Testing*, **13**(5), 43.

[26] Enemoto, S. and Mori, T. (1959) *Rep. Gov. Inst. Nagoya*, **8**, 418.

[27] Enemoto, S. and Furita, T. (1959) *Rep. Gov. Inst. Nagoya*, **8**, 428.

[28] British Standards Institution (1976) *BS 5288*.

[29] British Standards Institution (1978) *BS 5650*.

[30] International Standards Organisation (1978) *ISO 3999*.

[31] British Patents Office (1985) *BP.2159378A*.

5

Recording of radiation

5.1 INTRODUCTION

When X-rays pass through matter they are absorbed to an extent which depends on the thickness and density of the material and on the wavelength of the radiation. The properties of differential attenuation and the essentially linear propagation of X-rays provide the fundamental bases of radiography. The presence of internal cavities and inhomogeneities in an object will produce local variations in the spatial intensity of the emergent beam of an initially homogeneous beam of X-rays, in other words, the emergent beam will contain an 'image' of the internal structure of the object. However, X-rays are not directly perceptible to the human senses and the information contained in the X-ray beam must be converted to some form which can be appreciated by eye. A detecting medium which will reveal this information by means of a secondary interaction of X-rays with matter is therefore necessary. Such a detecting medium may give an indication of the spatial variations in intensity over an area, i.e. a two-dimensional detector, or it may be necessary or desirable to obtain a point-by-point measure of radiation intensity by a scanning procedure.

Several means of detecting X-rays are currently in use in radiology. The first and the most widely used method is the photographic effect. In this method the information is recorded as a variation in the silver deposit on a processed film and a full-sized permanent record, the radiograph, is produced. In general, film provides a comparison of spatial variations in X-ray intensity, but its great advantage is that it is an integrating device, and by extending the exposure time very low radiation fluxes can be recorded.

A second method of radiation detection is to observe the visible fluorescent light which is emitted by certain substances under the influence of X-rays. This effect is utilized in the well-known fluorescent screen on which the information is converted to visible form. However, the fluorescent screen does not produce a permanent record nor is it an integrating device, since the response of the screen is dependent on the intensity of the radiation and not on the quantity.

An extension of the fluorescent screen detector is to use a screen as a primary converter (X-rays-to-light), then use a closed-circuit television camera to view this screen, and convert the TV signal via an analogue-to-digital converter into a computer-stored signal. This technique is becoming increasingly important (Chapter 11).

X-rays possess the property of ionizing gases and the ions produced render the gas conductive. The existence of the ions, and hence the presence of radiation, can therefore be detected electrically. Ionization methods are capable of extremely high sensitivity and provide absolute values of X-ray intensity, but normally ionization chambers only give a point-by-point measurement across the X-ray beam.

In addition to ionization chambers there are several other forms of detector which convert an X-ray beam into an electrical signal; Geiger–Müller (GM) counters, scintillation counters, semiconductors, photo-conductive devices, etc. Most of these are point detectors, but some have recently been built as matrices and will be discussed later. Some of these detectors are also capable of transforming the X-ray beam into its different wavelengths.

5.2 PHOTOGRAPHIC EFFECT

Production and nature of the latent image

A photographic emulsion consists of a suspension of silver halide grains, usually chloride or bromide, suspended in a thin layer of gelatin. For handling convenience the emulsion is attached to a support which may be cellulose triacetate or polyester, or paper or glass, according to whether film, paper or plate is required. When radiation such as light or X-rays falls on the emulsion, extremely small particles in the silver halide crystals are converted into metallic silver. These traces of silver are so minute that the light-sensitive layer is, to all appearances, unchanged, but the number of particles formed is proportional to the quantity of light (or radiation) incident in the area, so that there is a 'latent image' in the emulsion.

When the film is placed in a developer, which is a weak solution of a reducing agent, the latent image becomes developed through the reduction of further silver halide and a stable, visible, black silver deposit is formed. The developer attacks the exposed silver grains differentially, grains which have received a high exposure are reduced quickly while those which received slight exposure are reduced or developed slowly. Once initiated the development of the individual grain continues until all the silver halide is reduced to metallic silver. In practice the action of the developer is continued for a limited time only; development is then stopped and all undeveloped halide is removed by means of a fixing agent. The amount of silver deposited then bears a definite relationship to the exposure, i.e. to the quantity of radiation incident on the film. Photographic developers are aqueous solutions which, in addition to developing agents, such as metol, hydroquinone, phenidone etc., contain accelerators, preservatives and restrainers. The accelerator is an alkali to speed up the developing process; the restrainer, usually potassium bromide, prevents the developing agent acting on the unexposed silver halide grains. Certain differences exist between the response of a photographic emulsion to visible light and to X-rays, and these are attributable to the much greater energy in the X-ray quantum. The most important difference lies in the resulting developability of the halide grains. In the case of light generally, one silver atom only is produced per incident quantum and initiation of development requires the absorption of several light quanta per halide grain. With X-rays one quantum alone is sufficient to ensure that a grain will be developed. This is because, on absorption of the X-ray quantum, the electron is emitted from the halide ion with a great deal of energy and it is therefore able to liberate secondary electrons from other halide grains, thus ensuring the reduction of a large number of silver ions to metallic silver. The increased certainty of initiation of a grain with X-rays explains certain differences between the characteristic curves for an emulsion on exposure to light and X-rays, and also shows why the response to X-rays does not exhibit reciprocity law failure.

X-ray film and its construction

An X-ray film consists of a thin base which is transparent and flexible, coated with a sensitive emulsion. The emulsion is attached to the base by means of a thin adhesive layer and is coated with an additional layer of gelatin (the supercoat) to minimize abrasion. To obtain as effective an absorption as possible X-ray emulsions are considerably thicker than those used for photography with light, and also there is a coating of emulsion on each side of the base.

Only a very small fraction of the X-radiation falling on the film is actually absorbed. If, however, a foil of metal such as lead is placed in intimate contact with the emulsion, the electrons ejected from the foil by the action of X-rays can also enter the emulsion and assist in the formation of the latent image. This is the principle of metal intensifying screens.

Photographic density

Before it is possible to consider the characteristics of photographic materials, an understanding of the meaning of the term 'density', as used in photography, is essential. In this context the term is used to mean 'film density' or 'photographic density'. Suppose a film of uniform blackness, that is, bearing a uniform silver deposit, is illuminated on one side by a beam of light of intensity I_0 and that the transmitted light intensity is I_1. The photographic density D of the film is defined as

$$D = \log_{10}(I_0/I_1) \qquad (5.1)$$

The ratio (I_0/I_1) is called the optical opacity of the film, and the reciprocal of the opacity is termed the transmittance. Thus, a film which transmits one-half of the incident light has an opacity of two and its density is 0.3. Similarly, films which transmit one-tenth, one-hundredth or one-thousandth of the incident light have densities of 1, 2, and 3, respectively. The instruments used to measure density are called densitometers.

As the silver deposit is formed of individual grains, the transmitted light is scattered and the spatial intensity of the transmitted light depends greatly on such factors as the distribution and size of the grains and on the condition of the incident light.

The measured density of a film will therefore depend on the optics of the densitometer. Three types of density are encountered in practice. 'Diffuse' density is measured when the incident light falls normally on the film and all the transmitted light is measured, or in accordance with the principle of optical reversibility, when the incident light is completely diffuse and only the normal or specularly transmitted light is measured. 'Specular' density is measured when the incident light is normal to the film and the normal component of the transmitted light is measured. When the incident light is completely diffuse and all the transmitted light is collected, 'totally diffuse' density is obtained. Diffuse density is usually used in practice.

Densitometers usually use a photocell to measure the amount of light transmitted through a film, and many such instruments use an electronic amplifying circuit. These must be stable over time and must cover a large range of light intensity, up to $10^4:1$ for a density of four.

Characteristic curve and its features

If the density of a processed photographic film is plotted against the common (base 10) logarithm of the exposure, whether this is to light or X-rays, a curve known as the 'characteristic curve' of the film, for the chosen conditions of processing, is obtained. Exposure E is the dose of radiation incident on the emulsion, i.e. the intensity I multiplied by time:

$$E = It \qquad (5.2)$$

In the case of a photographic emulsion exposed directly to X-rays, this reciprocity between intensity and time is strictly obeyed, but with exposures to light there can be reciprocity law failure and $E \neq It$ for any value of I or t.

The characteristic curve of density D against $\log_{10} E$ with X-rays exhibits certain well-defined and important features which are illustrated in Fig. 5.1. First it is evident that even with no exposure a small density is produced on development. This is the fog density and is composed of two factors: the inherent density of the base of the film, since this is not completely transparent, and a chemical fog density due to the fact that some grains are capable of being developed even without exposure. The actual fog density varies with the type and age of the emulsion and with the conditions of development, but a typical value for normal development for an X-ray film is between 0.2 and 0.3. As the exposure is increased from zero, density increases very slowly in what is usually called the toe of the curve; the toe is essentially a consequence of the

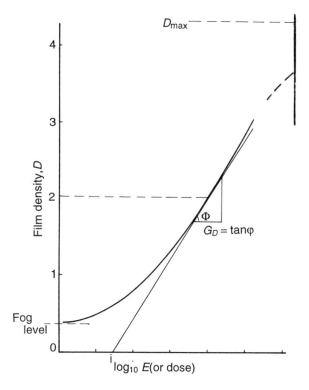

Fig. 5.1 Film characteristic curve for film exposed to X-rays, showing measurement of film gradient G_D at film density 2.0

logarithmic scale used for exposure. The curve then passes into a region where density varies approximately linearly with the logarithm of the exposure. As the exposure is increased still further the curve exhibits a shoulder, which is simply the consequence of a nearly complete chemical reaction. Eventually a maximum density D_{max} is reached. The curve finally passes into the solarization region where density decreases with increased exposure.

In the linear region, the curve has the equation

$$D = \gamma(\log_{10} E - \log_{10} i) \tag{5.3}$$

where γ is the slope of the approximately straight portion and $\log_{10} i$ is the intercept of the extrapolation of this portion on to the fog-density level. Sometimes i is used as a measure of the speed of the film. The slope G of the tangent to the curve at any point, i.e. the gradient of the curve often simply called the film contrast, (measured at density D) is given by

$$G_D = \frac{dD}{d(\log_{10} E)} \tag{5.4}$$

G_D varies with density and, although with visible light is often taken as constant and equal to γ over the approximately linear portion of the curve, there is good evidence that with X-rays there is no truly linear portion and G varies continuously with both exposure E and density D. G_D is therefore a measure of the film contrast available at the density used and is a very important fundamental parameter in radiography.

Owing to experimental difficulties in measuring G_D it has become fashionable to quote 'average gradients' for radiographic films, e.g. the average gradient between densities 1.5 and 3.5, which can be obtained by joining two points on the characteristic curve by a straight line, $G_{1.5-3.5}$ or $\bar{G}$. Sometimes the density values are taken as 'above fog density' (ISO: 7004).

Figure 5.2 shows curves of G_D against density for three different types of radiographic films (this is discussed later in this section). The maximum density D_{max} is obtained when all the silver halide is reduced to silver. This maximum is proportional to the weight of silver per unit area, which for ordinary photographic emulsions is such that D_{max} is of the order of three, but for double-coated radiographic films of the non-screen(direct) type, it is of the order of 12–16. Except for the emulsions intended for use with salt intensifying screens, where response is primarily due to light, there is no region where G is constant and it is common to find (especially with fine-grain emulsions) that G_D increases continually with density. There is no evidence that the use of metal intensifying screens has any effect on the shape of the film characteristic curve. This variation of G_D with density clearly indicates the importance

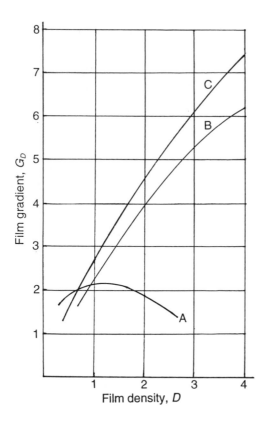

Fig. 5.2 Film gradient plotted against film density for various types of radio-graphic films. Curve A = salt screen film, used with salt screens; curve B = medium-speed film, used with metal screens or without screens; curve C = fine-grain film, used with metal screens or without screens.

of using G_D as a measure of film contrast, rather than γ, when considering radiographic films.

Factors affecting the characteristics curve

The shape of the characteristic curve for a given emulsion depends on the development process, the time and temperature of developer, the composition of the developing solution and the manner in which the development is carried out. Figure 5.3 shows a family of curves for a general-purpose radiographic film exposed under identical conditions, but given different development times in the same developer. It will be seen that increasing development time increases both the working speed and contrast of the film. A typical X-ray developer is designed to give

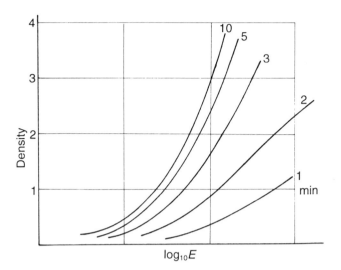

Fig. 5.3 Characteristic curves for radiographic films for different development times in typical radiographic developer at 20°C.

maximum speed and contrast without an undue increase in fog density. Historically, most X-ray developers are designed to work at a development time of 4–5 min at 20°C, but the advent of automatic film processors has led to processing at higher temperatures with special solutions designed to work at up to 30°C.

Film types

Films which are used for industrial radiography can be divided broadly into three main types.

1. Non-screen or direct-type films, i.e. films designed for use with metal foil intensifying screens, or without any screens.
2. Screen-type films, i.e. films designed for use with calcium tungstate or similar salt intensifying screens. In medical radiography these have now been subdivided to include films specially sensitized to the radiation from gadolinium oxysulphide intensifying screens (see also the discussion of fluorometallic screens in Chapter 7).
3. Non-X-ray films, i.e. films, usually of the single emulsion type, which are used for special purposes with X-rays, but which are not primarily radiographic films.

A screen-type film can be used with metal foil screens instead of salt screens, and it usually yields a lower contrast than a similar speed non-screen film. Used with metal screens, it will have a different characteristic curve to that obtained with salt intensifying screens. Many attempts have been made to group radiographic films according to some measure of their performance, as most film manufacturers market several non-screen type films of different speeds. The simplest grouping was: fast film; medium-speed, slow (fine-grain); very slow (very fine-grain). It was recognized that the slower the speed the finer would be the graininess of the image, and by implication the better the image detail.

Both ISO and CEN have attempted to put film classification on to a quantitative basis and the respective standards include:

1. a standard procedure for producing a characteristic curve;
2. a method of measuring and specifying film speed from this curve in terms of the X-ray dose required to give a film density of 2.0;
3. a measure of G_D or of average gradient;
4. a method of measuring granularity by linear scanning of a film with a uniform density of 2.0 with a 100 μm scanning spot and determining the standard deviation σ_D;
5. a film classification according to the value of G/σ_D for specific processing systems.

Both standards point out that the classification is for a film plus a processing procedure, not for a film alone. At present, one standard gives six classes, the other four classes, and a third proposal is simply to give a 'film quality number' according to the value of G/σ_D, where G may be a specific density gradient or an average gradient. To date the work excludes salt screen films. This method of film classification is basically a simplified version of the film quality index described in Chapter 8.

Farnell and Broadhead [1] showed that over a very wide range of radiographic films the granularity obtained with a particular energy of radiation is very closely proportional to the effective film speed, so that if films are rated according to their speeds this is also a rating of comparative granularity. Farnell and Broadhead measured granularity in a way that emphasizes the lower frequencies to which the eye is most sensitive, and the measurements correlated well with visual impression of graininess. This relationship would probably not hold for films with a markedly different silver/gelatin ratio (Chapter 8).

It is more difficult to generalize about contrast/gradient characteristics because of the effects of different processing solutions and techniques. The use of high-speed automatic processors, some of which work at high temperatures, has made possible much larger variations in contrast than when conventional 20°C solutions were almost universally used.

CEN: prEN–584–1: 1995 proposes that radiographic non-screen film systems (film and processing) be classified into six groups, based on

Table 5.1 Film classification, limiting values (prEN–584–1: 1994)

Film system class	$\sigma_{D(max)}$	$G/\sigma_{D\,(min)}$	G_{min} at $D = 2$ above fog	G_{min} at $D = 4$ above fog
C.1	0.018	300	4.5	7.5
C.2	0.018	270	4.3	7.4
C.3	0.023	180	4.1	6.8
C.4	0.028	150	4.1	6.8
C.5	0.032	120	3.8	6.4
C.6	0.039	100	3.5	5.0

measurements of minimum contrast gradient at density 2.0 and 4.0 above fog density and granularity (measured with a 100 μm scanning spot on a film of density 2.0). The relevant values are given in Table 5.1. In the four-group system, C.1 and C.2 are combined into one group, T.1; C.3 and C.4 become groups T.2, T.3; C.5 and C.6 become group T.4.

Table 5.2 shows film speeds and class groups of some of the most widely used commercial films. These data are not definitive, as they depend on the specific film developer employed, but they are based on commercial information available at the time of writing.

Effects of processing on the film characteristics

Automatic processing can effectively increase the film speed by as much as 50% compared with conventional processing at 20°C. It is possible for a film to move into another group when different processing is used, and similar effects can be obtained by short-time or extended-time development, so that film classification must take account of the processing system used.

Effect of agitation on uniformity of development (manual processing)

Lack of uniformity in the development process is manifest in two ways. First, in variations in overall density between different films which have had the same exposure, etc. Secondly, in local variation in density on the same sheet of film which has received a uniform exposure over its whole area. This second effect may not be very apparent when the film is examined by eye, but may be found to be quite severe when the film is scanned in a densitometer. Even with continuous agitation the films may still be of uneven density, as for example if the agitation merely consists of a slow lowering and raising of the film in the developer.

It is suggested that for minimum unevenness of film density the following schedule of agitation should be followed. The film should be given a circular motion in its own plane, continually for the first minute of development followed by alternate periods of agitation and

Table 5.2 Data on non-screen (direct type) industrial radiographic films

Film		CEN speed	CEN class
Agfa-Gevaert (Structurix)			
	D.2	50–30	C.1
	D.3	–	C.2
	D.4	100	C.3
	D.5	200	C.4
	D.7	400–250	C.5
	D.8	–	C.6
Kodak	M	100–64	C.2
	MX	125–100	C.3
	T	200–160	C.4
	AA	320–250	C.5/6
	AX	320–250	C.5
	CX	500–400	C.6
DuPont	30	–	C.1
	35	–	C.1
	45	–	C.2
	55	125	C.3
	65	250	C.4
	70	–	C.5
	75	500–400	C.6
Fuji	IX 25	–	C.1
	50	–	C.2
	80	–	C.4
	100	–	C.5
	150	–	C.6

Notes: Speed is 100 for $\log_{10} K_S = -2.05$, where K_S is the dose in greys for film density 2.0.
Conditions: 220 kV X-rays 8 mm Cu filter (HVT = 3.5 ± 0.2 mm Cu); lead intensifying screens 0.02 mm and 0.04 mm thick.
Where a range of speeds is given, this corresponds to two different processing systems, one manual and one automatic.

quiescence for 30 s at a time. This scheme has been found to give good results in most cases. However, some films, though developed uniformly up to the end of the time of development, increase in density so rapidly that the subsequent processing treatment, if not suitable, can allow slight development to proceed in an uneven manner in the emulsion before the fixing solution has had time to 'kill' the developer. It has been found that vigorous agitation in a spray rinser or a stop bath for about 20 s after development reduces this effect and assists in eliminating patchiness. The use of a nitrogen bubble agitator in the developing tank is finding increasing favour in radiographic processing units. This device produces short bursts of small gas bubbles at the bottom of the developing tank which, in rising, agitate the developer. The length of burst and the interval between bursts is preset and can be controlled electronically. Nitrogen gas is used to minimize oxidation of the developer. With automatic processors these difficulties of correct agitation procedures do not arise.

Developer replenishment

Replenishment of developer solutions is now a well-known technique for maintaining the activity of the developer and its height in the tank to a satisfactory working level. The purpose of the replenishment is to replace the small amount of developer that is carried over by the film to the stop bath. A modified developer solution is used as replenisher, so that both the total volume of developer and its activity are maintained constant. When a replenisher is added at the correct rate and films are allowed to drain for a standard time at the end of development, it has been found that the activity of the developer can be maintained accurately over a long period. Typically, the makers recommend that replenisher is used until the original developer solution has been replenished with an equal volume of replenisher.

Effect of radiation energy on film classification

If the characteristic curves of non-screen films obtained with different X-ray energies are compared, there is very little evidence for any change in shape with radiation energy within the wide range of X-ray and gamma-energies used in industrial radiography from, say, 50 keV to 30 MeV. However, the film speed and film granularity will change with X-ray energy, and ISO Standard ISO-7004 proposes that the film speed and classification group be measured with a range of radiations from 120 kV X-rays to Co-60 gamma-rays.

Film processing controls

Film processing systems can be controlled by the use of calibrated pre-exposed strips of film. The user processes these and measures the densities obtained. CEN Standard prEN-584-2: 1994 describes the production and calibration of such strips.

Viewing the film

When photographic film is used as a means of recording X-rays, the limitations imposed by the viewing conditions in which the film is examined must be considered. It is obviously no use to have information recorded on the film which is not detectable in the conditions under which the film is examined. The ability of the eye to discern both small detail and low contrasts depends on the luminance of the light reaching the eye and begins to deteriorate rapidly at luminances below about 30 cd m^{-2}. (Luminance is the correct term for what in normal English usage would be called 'brightness'; in photometry 'brightness' is reserved for the description of the saturation level of a colour.) Thus the intensity of illumination transmitted through a radiograph on the viewing screen should be above this minimum value, for optimum viewing conditions.

A film of average density 1.0 requires an incident luminance of ten times this value, a film of density 2.0 requires 100 times, etc., so that for films of density 2.0 a screen luminance of at least 3000 cd m^{-2} is required if the eye is to 'see' all the detail on the film.

Good viewing conditions require the elimination of extraneous light around the film or part of the film being examined; if not masked off this can cause glare. It is also necessary to eliminate extraneous light which may be reflected off the surface of the film into the eye. Such light reduces the effective contrast of the image and makes the minimum discernible contrast larger. Film viewing conditions are discussed in more detail in Chapter 7.

Image definition on the film

The smallest image detail that can be recorded on a film will depend, among other things, on the size and distribution of the developed silver grains in the emulsion. In general, the smaller the grains the finer the detail which can be resolved.

Grain size is a fundamental property of the emulsion and optical photomicrographs show that the developed grains occupy approximately the same positions as the original halide grains but, whereas the latter have definite geometric shapes and are too small to be visible with the naked eye, the silver grains after development have a filamentary structure. When a film is viewed a subjective impression of non-uniformity of silver deposit is obtained; this subjective impression is called 'graininess', and is due to the uneven distribution of developed grains or clumps of grains, i.e. the statistical variation in the number of grains per unit area of viewed film, and by the overlapping of the clumps of grains. In general, the smaller the grain of the emulsion the smaller the impression of graininess. The graininess of an emulsion, however, depends on a number of factors other than the type of emulsion. Graininess increases with the effective energy of the radiation, and a radiograph taken with gamma-rays is usually more grainy than one taken with X-rays. Graininess also increases with the degree of development and is affected by the composition of the developer, although special fine-grain developers are rarely, if ever, used in radiography because of the loss in contrast which their use causes.

There is a second characteristic of a photographic emulsion exposed to X-rays which acts as a limit on definition. This is the effect of secondary electrons, released on absorption of a quantum of X-ray energy, in sensitizing silver grains adjacent to the grain initially absorbing the primary quantum. Thus a small string of grains may be rendered developable by one X-ray quantum and the effect is to produce a blurring of image detail; two-dimensionally, the effect is to produce a disc image on the film instead of a point image. This effect is usually known as film or inherent unsharpness U_f. When a quantum of X-radiation is absorbed in a halide grain the grain is rendered developable and at the same time a photoelectron is released. With high energy X-ray quanta having energies greater than is required for ionization, the excess energy appears as kinetic energy of the electron. This therefore may have sufficient energy to reach an adjacent silver halide grain while still having sufficient energy on being absorbed to render this second grain developable. The higher the quantum energy of the incident radiation the greater will be the range of the electrons; a single quantum of X-rays, instead of affecting only a single halide grain can, if sufficiently energetic, produce a small volume of developed silver in the emulsion. Bromley [2] and Herz [3] showed that with X-ray energies up to 33 keV, only one halide grain becomes developable for each absorbed quantum, but that as the incident energy increases, the number of grains per quantum also increases. With 1000 kV X-rays they estimated that about 80 grains are made developable for each quantum absorbed in a typical X-ray film emulsion.

Considering the image of a sharp metal edge, instead of a theoretically sharp image, as shown in Fig. 5.4(a), the density variation across the image of the edge will be of the general shape of that shown in Fig. 5.4(b), the width of the band of density change being taken as a measure of the film unsharpness. The curve obtained experimentally is usually similar to Fig. 5.4(c), which is 'smoothed' to the shape shown in Fig. 5.4(d).

In practice, if X-ray film is used between metal foil or with salt intensifying screens, the individual effects of film and screen unsharpness cannot be separated, and the term 'inherent unsharpness' is sometimes used to describe the effective total unsharpness of the film–screen combination. With suitable densitometric equipment it is possible to measure film unsharpness directly by making a radiograph of a suitably sharp metal edge placed close to the film emulsion and accurately aligned with the X-ray beam, the source of radiation being at such a distance that the geometric unsharpness is kept very small. The variation in density on the film across the image of the metal edge is then measured with a microdensitometer, and a curve of the variation in density against distance is plotted.

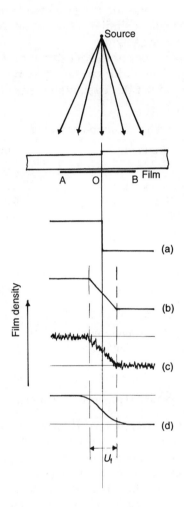

Fig. 5.4 Measurement of inherent (film) unsharpness U_f.

Experimentally, several precautions are necessary if the results are to be accurate and of value for practical radiography.

1. As values of unsharpness are frequently used to calculate radiographic sensitivity values, the influence of unsharpness on the image of small low-contrast detail is of more practical importance than its

influence on a high-contrast image. Thus, the density change across the image should be made as small as possible.

2. In industrial radiography the radiation reaching the film is heavily filtered by the specimen: similar radiation should therefore be used for unsharpness measurements.

3. To obtain accurate curves of the density variations across the image of the metal edge, the width of the scanning spot (or line) of the microdensitometer must be considerably smaller than the unsharpness to be measured.

Table 5.3 Film unsharpness values

Radiation (filtered)	U_f(mm)
50 kV X-rays	0.03
100 kV X-rays	0.05
200 kV X-rays	0.09
300 kV X-rays	0.12
400 kV X-rays	0.15
1 MV X-rays	0.24
2 MV X-rays	0.32
5.5 MV X-rays	0.46
8 MV X-rays	0.60
18 MV X-rays	0.80
31 MV X-rays	0.97
Iridium-192 gamma-rays	0.17
Caesium-137 gamma-rays	0.28
Cobalt-60 gamma-rays	0.35
Ytterbium-169 gamma-rays	0.07–0.10[*]

[*] Value depends on filter/specimen thickness.

Measured film unsharpness values [4] are given in Table 5.3 for a range of radiation energies. In each case filtered radiation has been used, with a filter thickness corresponding broadly to the typical specimen thickness for that radiation (see also section 6.3). The values in Table 5.3 are also plotted in Fig. 5.5 and show the continuous rise in U_f with radiation energy. Very little difference in U_f for a particular X-ray energy was found for different radiographic films, and this confirms that most films have a similar silver/gelatin ratio. There are other methods of specifying film unsharpness, rather than the simple measure of U_f used here (Chapter 6). Alternative methods of image definition measurement are discussed in Chapter 8.

Graininess: granularity

The third factor likely to affect the discernibility of fine detail in the image is that the image is built up in the film emulsion of silver grains.

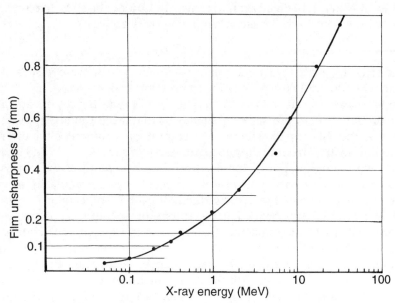

Fig. 5.5 Experimental curve of inherent (film) unsharpness for a fine-grain film plotted against X-ray energy (filtered radiation).

Common sense and experience, for example half-tone reproductions in newspapers and magazines represent typical examples, suggest that a coarse-grained image conveys less detail than one of finer grain; thus the grain size must have some influence on the amount of recorded detail.

It is also easy to demonstrate that graininess and unsharpness are not two names for the same effect; the image of an 'edge' can be blurred, without being granular. If one considers the case where geometric unsharpness is present, this intensity distribution represents an unsharp image; only the fact that the image is recorded on a photographic emulsion makes the final image grainy as well as showing the effect of radiographic unsharpness.

The concept of 'graininess' has been the subject of much confusion. In a photographic emulsion silver halide grains, usually of the order of $1 - 10\,\mu m$ in diameter, are distributed at random in a layer of gelatin. On exposure and development some of these grains are converted into blobs of silver and the distribution of these developed silver particles must also be random. In such a random distribution there is a natural statistical fluctuation in the number of grains per unit area across the emulsion. If it is assumed that in a small area A there are N grains each of average area a, then the density of this area will be proportional to $N \times a$. But if an adjacent area of the same size A is considered, the number of grains in this area will not necessarily be exactly N. The mean deviation in the value of N in a statistically random distribution can be shown to be $N^{1/2}$, so that the density of the second area may vary between

$$(N \pm N^{1/2})a \tag{5.5}$$

and it is this statistical fluctuation of density which the eye sees as it scans the film and which gives the impression of graininess.

If the scanning area is made larger N will also be larger, and the fluctuations are proportionately smaller so that the graininess appears to be less, and this applies whether the scanning instrument is the eye (by standing further from the film a greater area of film is subtended by the effective scanning spot of the eye and the impression of graininess is reduced) or an instrument (if the scanning spot of a granularity instrument is made larger, the density fluctuations become smaller). A great deal of experimental work has been done on scanning spot size with granularity meters to make the granularity measurements correlate with visually observed graininess.

Graininess is most frequently measured by estimating the blending distance, i.e. the viewing distance at which, under closely specified conditions, the observer no longer has any visual impression of a grain structure. Alternatively, the magnification at which a grain structure first becomes apparent for fixed viewing conditions may be used. Granularity has been defined in terms of the root-mean-square fluctuation in density, when a microdensitometer trace is made across a uniformly exposed and developed piece of film. If the instrument has a scanning spot of area a, the granularity G is defined as

$$G = \left(\frac{\text{mean square variation in density}}{\text{area of scanning spot}} \right)^{1/2} \tag{5.6}$$

and is usually expressed in density-micrometres.

Neither graininess nor granularity, as so far defined, are directly related in any simple manner to the amount of detail recorded on a radiograph, as defined in terms of contrast and unsharpness. Therefore, a different method of treating the problem of image detail, which allows granularity to be brought into consideration, is discussed in Chapter 8.

In attempting to obtain graininess measurements by blending distance on radiographs it was noted that a grainy image of uniform density has several features. First, there was a comparatively fine grain structure. In some films, particularly the slow-speed ones exposed to low-energy radiation, this pattern seemed to be reasonably uniform and the blending distance was fairly easy to determine. But on most films there appeared to be patches or strings, each appearing to consist of an agglomeration of a number of grains; these could still be discerned when the initial graininess had blended to a uniform density, and the blending distance usually quoted is when these strings or clumps are no longer discernible. The graininess seen also gave an impression of varying in contrast; on films exposed to higher energies the grain strings, although not much larger, were more prominent and this factor appears to be the prime cause of an increase in blending distance. In addition, on some radiographs there appears to be a low contrast coarse mottle, much coarser than the grain strings, and this is probably quantum mottle.

Using blending distance as a measure of graininess, by large size projection of a uniform density, the film graininess is found to rise gradually from a low value for visible light or very low energy X-rays (30 kV) to a maximum for around 1 MV X-rays, after which the rise is either very small or might even be negative. The maximum increase in blending distance is about 50%, which is very similar to the change shown in granularity measurements.

Intensifying screens

Even with the double-coated X-ray films currently available only a small fraction of the total X-ray energy reaching the film is absorbed by the film. A greater proportion of the energy can be made to contribute to the formation of an image by the use of intensifying screens. These may be either metal foil screens, where the intensification is by electron action, or salt screens where intensification is by fluorescent light (visible or ultra-violet).

The intensification of metal screens is due to the emission of electrons under the action of X-ray absorption, so intensification increases with the atomic number of the metal, and although it would be advantageous to use a metal with as high an atomic number as possible, other considerations such as cost and malleability usually restrict the choice to lead foil. Frequently the lead is alloyed with a few percent of antimony or bismuth to increase its hardness and to improve resistance to abrasion. The alloying metal must be evenly distributed through the lead and not be in a segregated form; occasionally metal intensifying screens produce a pattern on the film and this can usually be traced to a segregate or to lead oxide inclusions. The use of tantalum, copper and tungsten screens has been proposed, especially for high energy X-rays where they have some special advantages (Chapter 7), but for most purposes lead screens are used.

The thickness of lead screen is usually between 0.02–0.15 mm, and as the screens must be pressed tightly into contact with the film emulsion, they are usually mounted on a supporting card. Some manufacturers market film, particularly strip film for weld inspection, with built-in metal screens integral to the film envelope. These screens are very thin (0.02 mm) and are cemented to the inside surface of the paper envelope. The film is packed under vacuum to ensure good film-to-screen contact.

As intensifying screens are in close contact with the film emulsions, there is little opportunity for electrons generated in the screens to spread sideways and cause any blurring of the image, and except with very thick screens and high-energy radiation, the use of thin metal intensifying screens has only a very small effect in increasing the blurring of the image when compared with an image recorded without the use of screens.

With medium-energy X-rays, thin metal screens will produce an intensification factor of between 2 and 5, that is, the exposure for a chosen film density can be reduced by this factor.

With low energy X-rays only electrons from the surface of the screen nearest the film are emitted, the rest being absorbed in the lead, and the screens may have an intensification factor of less than one.

Another important result of the use of metal screens is the reduction of the amount of scattered radiation reaching the film, due to preferential absorption of the softer scattered radiation by the front screen. Even when the intensification factor is less than one, the front screen still has a useful function as a filter. Salt intensifying screens are rarely used in industrial radiography, because they produce a considerable loss of detail compared with radiographs taken with metal screens. Salt intensifying screens are extensively used in medical radiography, chiefly because their high intensification factor $(\times 30 - \times 100)$ permits a large reduction in the dose of radiation to the patient. As a secondary point, the comparatively low X-ray energies used in diagnostic medical radiography lead to very high-contrast images and this high contrast compensates, to some extent, for the very poor definition. Salt screens were used extensively in the early days of industrial radiography, when the output of radiation from both X-ray and gamma-ray sources was so low that there was no alternative if the exposure times were to be kept reasonably short, but their present day use is rare and limited to a few specialized applications such as flash radiography.

The poor definition obtained with salt screens is usually attributed to the general scattering of light in the fluorescent layer. This blurring can be measured as a screen unsharpness U_s, which is analogous to film unsharpness, and for most screens is of the order of 0.2–0.4 mm. This explanation of screen unsharpness is not entirely satisfactory and it seems probable that the unsharpness of a radiograph taken with salt screens can be related to two causes which have been called 'structure mottle' and 'quantum mottle'. The former is caused by structural inhomogeneities in the phosphor coating of the screens and the latter is caused by statistical fluctuations in the spatial distribution of the X-ray quanta absorbed in the screens. Both these effects are considerably larger than the effects of film graininess, and it is suggested that of the two causes, quantum mottle is much the more serious. Since the light produced per absorbed X-ray quantum in the intensifying screen increases with the quantum energy of the X-ray photon, fewer high-energy photons are necessary to produce a given film density; the quantum mottle may therefore be expected to be more pronounced with higher energy X-rays.

Other materials

Occasionally materials other than radiographic film are proposed for the recording of X- and gamma-radiation. Silver emulsions coated on a paper base were used extensively at one time for routine inspection of ordnance fuzes, where only a limited image density range is necessary to examine for specific constructional details. Generally, on paper the flaw sensitivity is poorer because of a lower contrast gradient and the expo-sure time has to be more accurately controlled than on film, but X-ray paper is much cheaper than film.

Photographic bromide paper used with a salt intensifying screen has also been marketed for rapid, cheap radiography, to detect large flaws in light-alloy castings with low energy X- rays. A Dry Silver film has been marketed for photography, which can be used with a fluorescent screen with X-rays. The image is developed by heat, generated by an electric current applied to a conductive backing layer on the film's polyester base. Development takes only a few seconds and a high $D_{(max)}$ is claimed, which should result in a reasonably good flaw sensitivity.

5.3 IONIZATION

Ionization processes [5]

On the simple concept of atomic structure, an atom consists of a central positively-charged nucleus in which practically all the mass of the atom is concentrated, surrounded by negatively-charged orbital electrons. The number of electrons is such that their combined charge balances the charge of the nucleus, so that a normal atom is electrically neutral. The electrons in orbits nearest the nucleus are most strongly held by the nuclear charge and require the greatest amount of energy to remove them from the influence of the nucleus. As the radius of the orbits increases the electrons become less and less tightly held and require decreasing amounts of energy for their removal. An electron, however, cannot move into any orbit, but only in those permitted by the require-ments of quantum theory, in other words, it can only move in orbits of fixed energy levels. An electron can move between orbits if the energy given to the electron is equal to or greater than the difference in energy between the two levels. In the normal state of the atom the electrons occupy the lowest energy levels, i.e. the orbits nearest the nucleus, but there exist beyond the filled orbits several to which the electron can move if given the right amount of energy. An atom with an electron which has transferred to one of these higher energy levels is said to be 'excited'. Generally the electron does not remain for very long at the higher energy level, but quickly drops back to its former level with the emission of the extra energy in the form of radiation. If the electron originally came from one of the outer occupied orbits the emitted radiation will be in the visible or ultra-violet spectrum, and this is the phenomenon of 'fluorescence'. If the electron originated from an orbit near the nucleus then, because of the greater difference in energy of the two levels, the radiation emitted is of much shorter wavelength. This is the characteristic X-radiation.

If an electron absorbs sufficient energy it may be removed completely from the atom, which is then left with a net positive charge of one unit. The atom is said to have become 'ionized' and an ion pair, consisting of a positive ion and an electron, has been formed. The excess of energy above the amount required to release the electron is imparted to the electron as kinetic energy. This kinetic energy may be sufficient for the electron to cause further ionization by collision with other atoms, or the electron may be emitted as a photoelectron.

In solids and liquids atoms do not exist very long in an ionized state. Eventually they pick up an electron and become neutral again, but in gases, especially gases at low pressure, where the individual atoms and molecules exert little influence on one another, positive ions and electrons may exist for a long time before recombination takes place. Thus, ionization of a gas leads to the formation of free charge which can be collected by an applied electric field.

Gaseous ionization may be produced either by fast-moving particles such as alpha- or beta-rays or by electromagnetic radiation such as X-rays or gamma-rays. When a particle moves through a gas it dissipates its energy by ionizing the atoms with which it collides and the specific ionization of the particle is defined as the number of ion pairs produced per centimetre of its path. The total number of ion pairs produced is dependent on the initial energy of the particle. The passage of X-rays or gamma-rays through a gas causes the ejection of high-energy electrons from the atoms and these primary electrons move through the gas producing ion pairs. The specific ionization of the primary electron is dependent on the energy absorbed from the incident radiation. Uncharged particles such as neutrons can be detected indirectly by ionization produced from breakdown products, such as protons, electrons or alpha-particles, which follow the capture of the neutron after collision with a nucleus.

Almost any gas will ionize when irradiated, but not all gases are suitable for use in an ionization chamber. The gas should have a low saturation potential, i.e. one in which recombination takes place at a low potential, and the gas atoms should be such that the free electrons cannot readily become attached to form negative ions. In other words, the properties of the gas should be such that every electron produced by the event is collected at the electrodes under the applied field and is not lost by recombination. For fast counting the transit-time for free electrons should be as short as possible, and sometimes a small quantity of a second gas is added for this purpose.

The esential component of an ionization chamber is the envelope in which the ionizing gas is contained, in which there are two electrodes between which a potential difference can be applied. One electrode can be the chamber wall. If the potential is produced by an electrostatic charge, or by momentary connection of the electrodes to a potential source, the instrument is usually known as an electroscope and behaves as a leaky capacitor when exposed to radiation. The simplest form of electroscope is the well-known gold-leaf electroscope. Electroscopes of this type are low-capacity instruments and the total charge contained in them is small. They are not capable of measuring large doses of radiation, but since they integrate the dose over a period of time, very small dose-rates can be measured.

Small-capacitor ionization chambers are also used for the recording and measurement of small total doses of X-radiation. These consist of a chamber in which the central electrode is very efficiently insulated so that they can be charged to a few hundred volts, and have a negligible leakage over a period of a few days, in the absence of radiation. In the presence of ionizing radiation there is a loss of potential due to ionization in the chamber, and the quantity of radiation received is proportional to the loss of charge. Typical maximum doses for this type of ionization chamber are 0.2 to 50 R, or 5×10^{-5} to 1.3×10^{-3} C kg^{-1}. A 1 VR^{-1} chamber will have a volume of about 1 cm^{-3}. The measurement of charge in these chambers, which have a very small capacitance, requires special apparatus since the capacitance of the measuring instrument is connected in parallel with the chamber and there is thus normally a large drop in voltage.

Ionization chambers with attached heads

Instead of relying for charge storage on its own electrical capacitance, an ionization chamber is very frequently connected to an independent voltage source and measuring system. The latter can be integrating or instantaneous. If a long cable is used, the time-constant may become inconveniently long at low exposure rates. One way of overcoming this difficulty is to place an electrometer tube close to the ionization chamber, or to reduce the voltage charge on the cable by applying feedback from the amplifier. The active volume of an ionization chamber is small and

the cable connecting the chamber with the measuring system must therefore be quite free from air spaces, the ionization of which could cause
stray currents in the collector lead; particular care is necessary at cable
junctions.

The choice of material for the wall and gas of an ionization chamber is
primarily governed by the quantity to be measured. If the chamber is to
measure exposure, it should be made air-equivalent, i.e. it must have an
effective atomic number closely matching that of air. If it is to measure
absorbed dose in a particular medium, both walls and gas should be
matched to that medium. Materials are matched for a particular radiation if the absorption leads to the same flux density and energy
distribution of secondary ionizing particles.

When measuring low and medium energy X-rays up to an energy of a
few megavolts the ionization chamber can readily be made thick enough
for electron equilibrium to be attained. If the wall material is reasonably
air-equivalent such a chamber can then be calibrated to measure exposure in roentgens. For an X-ray beam of known energy it is advantageous
to choose a wall thickness near the flat maximum of the build-up curve,
since the influence of wall thickness is then minimal.

Current–voltage characteristics

If a potential difference is continuously maintained across the electrodes
of an ionization chamber, the characteristics of the chamber vary with
the voltage. Suppose an ionizing event occurs in the gas, leading to the
formation of ion pairs. The electrons are attracted to the positive plate
and the positive ions to the negative plate. Owing to their much smaller
mass the electrons are rapidly collected, while the heavy positive ions
remain relatively stationary. The removal of the electrons produces a
pulse of current, and for a given radiation flux the number of pulses
recorded depends on the applied voltage. The current from an ionization
chamber can be measured as

1. the average direct current
2. the number of pulses.

Initially, in the region which is known as the ionization region, the
number of pulses received per second is constant. Every event produces
a burst of electrons which, provided the voltage is high enough to sweep
them out of the volume before recombination can occur, are collected at
the anode. The voltage, which is just sufficient to collect every electron,
is known as the 'saturation voltage', and provided this voltage is exceeded the size of the pulse is also independent of voltage. The rate of
production of current pulses, however, is proportional to the radiation
flux. Under normal working conditions the incident flux is sufficiently

high that a continuous saturation current proportional to radiation intensity is produced.

This is the region in which ionization chambers operate. Using small ionization chambers of about $10–20 \, cm^3$, typical ionization currents range from 10^{-12} A for $1 \, mRh^{-1}$ to 10^{-9} A for $10 \, Rmin^{-1}$. As the voltage across the electrodes is increased, the proportional region of the pulse/voltage characteristic is entered. The speed of the electrons is now increased until they have sufficient energy to produce more electrons by further collisions with gas atoms. The process is obviously cumulative, and an electron from one initial ion pair is able to bring about a large increase of ions (known as a Townsend avalanche). Both the pulse height and the number of pulses per unit time steadily increase with increasing voltage. For a given working voltage the pulse is proportional both to the number of ions produced by the initial event and to the gas amplification factor, M, i.e. the number of electrons freed by the collisions, M, varies from unity in the ionization region to about 10^4 at the upper end of the proportional region. The initial number of ions produced depends on the type of radiation or particle entering the chamber, i.e. on the specific ionization of the initial event, and a proportional counter is therefore able to discriminate between different incoming events. Owing to the higher ionization currents involved a proportional counter provides stronger signals which require less external amplification. They are also faster than ionization chambers in response, which means that higher counting rates are possible.

Geiger counters

With still further increase of voltage the number of pulses per second continues to increase, but proportionality between pulse height and amplification factor is no longer maintained. Eventually a second region is reached where independence exists between the number of pulses per second and voltage. This is the Geiger region, which is of great importance. Due to the high voltage the discharge is no longer confined to a small region of the gas but spreads to fill the whole volume between the electrodes. It therefore follows that for a given voltage on the chamber, the pulse size is constant, i.e. it is independent of the number of ions produced by the initial event, though the absolute size of the pulse continues to increase with voltage, and it is only the total number of pulses that gives a measure of the radiation intensity. The output pulse is very large (1–50 V) and can be recorded with little or no amplification. The flat portion of the curve is called the plateau, though in fact it is not strictly flat and has a slight upward slope due largely to end-effects in the chamber. Beyond the plateau the number of pulses per second increases rapidly and the counter goes into a semicontinuous discharge.

Under the influence of the high voltage within the Geiger region the electrons are quickly collected, leaving the positive ions occupying the active volume and moving slowly to the cathode. The presence of the positive ions has important consequences in the functioning of the Geiger, or Geiger–Müller (GM), counter. If their velocity on arrival at the cathode is high enough, they too can produce secondary electron emission which tends to maintain the discharge. This tendency to continuous discharge must be prevented or quenched, and quenching can be carried out in two ways. The first is to employ an external circuit which, immediately the pulse is recorded, lowers the voltage across the counter below the value required for the positive ions to liberate electrons at the cathode, until all the ions are removed from the gas.

An alternative method of quenching the discharge is by introducing a suitable polyatomic gas into the chamber. The counter then becomes self-quenching, due to a somewhat complicated mechanism dependent on dissociation of the polyatomic molecules. Self-quenching counters have a definite, limited lifetime of the order of 10^9 counts. A common quenching gas is a 10/90 (%) alcohol/argon mixture at a total pressure of 10 cm Hg. The halogen gases also exhibit self-quenching properties. Counters employing halogens must be constructed with stainless-steel electrodes which are not corroded by the gas.

In summary, therefore, gas-filled counting tubes have been among the most widely used of radiation detectors, and although for many purposes they have been superseded by scintillation counters and are subject to growing competition from solid-state devices, they may be expected to retain an important place in radiation dosimetry. Counting tubes are not ideal; their response is more closely related to the number of ionizing particles entering the tube than to the energy absorbed. The GM tube delivers a virtually identical pulse for any directly ionizing particle so long as it generates one primary ion pair, so that interpretation of an observed number of counts in terms of absorbed dose must involve some knowledge of the nature and energy of the radiation. GM counters are very simple devices and are well-suited to the construction of portable instruments for radiation monitoring, where high accuracy is not required.

The proportional counter comes much nearer to direct measurement of dose; it delivers a pulse proportional in height to the number of primary ions formed in the gas, and so proportional also to the energy dissipated in the gas. Any GM tube can be operated in the proportional region; there is no difference in construction. The pulses will be much smaller, so a higher gain amplifier will be needed, but this is offset by several advantages: there is a greater latitude of filling gas; the counter operates much faster; the proportional characteristics enable the pulse height distribution to be determined by scanning with a single channel analyser or by differentiating a curve of pulse rate against discriminator setting.

An adjustable high voltage supply to operate any counter is necessary; this may be as low as 300 V for a halogen tube or higher than 3000 V for some proportional counters; GM counters require 1% voltage stability, but proportional counters require 0.1%.

The individual events that trigger a counter occur independently and are therefore distributed randomly in time. If repeated observations are made over equal periods, in a constant radiation field, the results will show a scatter the dispersion of which is such that, if N is the mean number of counts observed, the standard deviation of N is given by δ, where $\delta = N^{1/2}$ or, expressed as a percentage, $100 \, N^{-1/2}$(%). There are, therefore, fixed probabilities that the true number of counts N_t lies between $(N \pm xN^{1/2})$, where $x = 1, 2, 3 \ldots$:

$$x = 1, \text{probability } P = 68.3\%$$

$$x = 2, \qquad\qquad P = 95.5\%$$

$$x = 3, \qquad\qquad P = 99.7\%$$

As N is proportional to the counting time, the precision of a measurement varies as the square root of the time taken to collect the data. For many applications these simple statistical ideas are sufficient, but more rigorous treatments are sometimes necessary.

5.4 SCINTILLATION COUNTERS

A scintillation counter consists of a phosphor which converts some of the energy of incident X-rays or gamma-rays etc. into light photons (scintillations), together with a photomultiplier tube which transforms each single light pulse, by an electrical cascade mechanism, into an electric pulse.

Many solid and liquid substances show the property of luminescence, or emission of visible light, when exposed to ionizing radiations such as X-rays, and the development of photomultiplier tubes has made it possible to record the light flashes that correspond to the absorption of single quanta in a luminescent material. The most important groups of scintillators are the organic crystals of the hydrocarbon type, the inorganic crystals and powders of the alkali halide and zinc sulphide types.

For high energy X-rays and gamma-rays thick crystals are usually used, thallium-activated sodium iodide being one of the most commonly used materials. Owing to the large mass of phosphor the detection efficiency can be much higher than with a Geiger or proportional counter. The phosphor must be transparent and all but a small area of the phosphor adjacent to the photomultiplier pick-up surface can be coated with a reflecting layer to prevent light loss. A scintillation counter is several orders of magnitude faster than any detector depending on gaseous ionization, and the speed of counting is only limited fundamentally by the afterglow of the phosphor. The counting rate may exceed 10^7 pulses s^{-1}. The sensitivity of the counter approaches ten times that of a Geiger counter, although to achieve this figure special precautions must be taken to reduce the dark current (noise) of the photomultiplier to a minimum.

As all the energy of an incident quantum can be absorbed and converted in a large scintillation crystal the light output is proportional to the energy of the quantum. Thus a scintillation counter can be used as an energy spectrum analyser.

Two methods have been employed for radiation measurements with scintillation detectors. In the first method, the scintillations produced by fast electrons are individually counted by means of conventional circuits for pulse amplification and analysis. In the second method, the average DC current corresponding to the total light output of the scintillator is measured. The fast response of scintillators with short decay times can only be utilized by pulse methods. If a simple scaler arrangement is employed each particle with more than a certain threshold energy will produce one count irrespective of its energy, but pulse discrimination circuits can be used to determine the energy distribution of the particles absorbed.

The DC output of the photomultiplier tube is a very precise measure of the total light output of the scintillator. Each light quantum originating in the scintillator and reaching the photocathode has approximately the same probability of ejecting an electron, and each photoelectron produces the same average charge on the anode; therefore the output current of the photomultiplier is proportional to the rate at which radiation energy is absorbed in the scintillator, provided that the optical coupling remains fixed.

The decay time of the light intensity does not conform to any simple mathematical relation. The decay time is defined as the duration of the

Table 5.4 Typical scintillators and their characteristics

Material	Density (gcm^{-3})	Atomic number	Wavelength emission (Å)	Relative light yield	Decay time (ns)
Anthracene	1.25	5.8	4450	1.00	25
Stilbene	1.16	5.7	4100	0.73	7
ZnS (Ag)	4.1	27	4500	2.0*	>1000
CdS (Ag)	4.8	44	7600	2.0*	>1000
NaI (Ti)	3.67	50	4100	2.0	250
KI (Li)	3.13	49	4100	0.8	>1000
CsI (Ti)	4.5	54	White	1.5	>1000
CaWO$_4$	6.1	59	4300	1.0	>1000

*Depends on treatment and intensity.

strongest initial light pulse or the time to $1/e$. In some cases, weak light emission may continue for several orders of magnitude longer than the decay times. Typical scintillators and their characteristics are shown in Table 5.4.

5.5 SEMICONDUCTOR DEVICES

Basic mechanisms

The electrical conductivity of certain crystals, such as cadmium sulphide and cadmium selenide, is very low ($R = 10^{10} - 10^{12}$ Ω) under normal conditions when a voltage is applied across the crystal. When irradiated with X-rays the conductivity is found to increase by nearly 10^4. This property of photoconductivity, so called because it was originally found to exist in certain crystals when irradiated with light, forms the basis of another method of recording and detecting X-rays.

These materials, called semiconductors, are a class of substance between insulators and conductors [6]. The imperfections in the crystal lattice, which act as electron traps, play an important part in the mechanism of photoconductivity. In addition to imperfections in the lattice, impurities such as atoms of other substances which can easily lose an electron are present as activators. When a potential is applied across the crystal in the absence of radiation the small dark-current which flows is due to electrons leaving their normal orbits and becoming free, through thermal impact from the activators. The number of electrons present in the crystal at each instant is determined by the stationary negative space-charge of the electrons held in the trapping states. If an electron leaves the crystal at the anode another enters at the cathode so that the total number of electrons remains constant. When an X-ray quantum is absorbed some of the crystal atoms are ionized and the liberated electrons are able to move freely in the crystal lattice. The ionized crystal atoms are quickly restored to the neutral state by electrons from the activators which now themselves have a net positive charge, which tends to compensate for the negative space-charge of the trapped electrons. A chain reaction is initiated and persists as long as the positive space-charge exists. The length of the electron chain is determined by the probability of the positively-charged activators recombining with the electrons.

If the field strength across the crystal is of the order of 500 V mm^{-1}, approximately $10^4 - 10^5$ electrons are carried across the crystal for every electron initially expelled from the crystal atom. Thus, the conductivity of the crystal is increased as long as it is irradiated, and the photocurrent increases with the radiation intensity. Due to thermal agitation, the distribution of the space charges is continually undergoing rearrangement, a process which is responsible for the delay in reaching the final current when a constant irradiation is applied or removed. This delay may be a fraction of a second, or up to several minutes with very low intensity radiation.

Cadmium sulphide crystals, for use as X-ray detectors, have the advantage of small size, comparative cheapness and robustness. A typical crystal would be 2×5 mm with a sensitive area of 10 mm^2; the crystal will have indium electrodes at both ends of the C-axis and be mounted in a light-proof cover.

The chief advantage of all solid-state conductivity devices is that their high density and low ionization potential enables dosemeters of very small volume to be constructed. If a voltage is applied to a photoconductor, a current can be measured which is a function of the absorbed dose input or exposure-rate at the photoconductor. A small photocurrent may of course be made to flow through an external circuit.

Semiconductor junction detectors (diodes)

The properties of semiconducting material are exploited by adding traces of impurity (one part in 10^8). There are two types of impurity, donor (n-type) and acceptor (p-type). If n-type material is introduced into the crystal lattice a surplus electron is made available which can easily be displaced. Similarly, a p-type impurity causes an electron deficit. A semiconductor-rectifier is made by adding n-type material at one end of a crystal and p-type at the other and heat-treating the crystal.

Junction detectors with reverse bias behave much like solid-state ionization chambers. The n-region of silicon conducts electricity because it contains a surplus of electrons. The p-junction is deficient in electrons and therefore conducts only by the movement of positive holes. Electron–hole pairs are generated by photoelectric and Compton absorption effects. If a potential is applied across the device in the direction such as to attract both types of charge carrier towards the junction between the regions, a current will flow. If the polarity is reversed so that the charge carriers of both types are attracted away from the junction, a depletion layer, devoid of charge carriers, is formed in the neighbourhood of the junction and only a small leakage current will flow across. The thickness of this layer is controlled by the applied potential, which is called the reverse bias. If ionizing radiation then generates new pairs of electrons and holes in the depletion region they will be collected by the potential across the layer, constituting an ionization current just as in a gaseous ionization chamber. Each particle absorbed gives rise to a pulse of charge, the size of which depends on the number of electrons or holes; the rise-time is $10^{-8} - 10^{-9}$ s. Radiation dose rates can be measured by p–n junctions without applied bias, and lower rates can be measured than with currents collected when bias is applied, by measuring the current through a very low impedance external circuit, or by measuring the open circuit potential difference generated across the junction. Typical values quoted are:

- depletion layer thickness $\quad$ 0.25–0.5 mm
- open-circuit photovoltage $\quad$ $10\,\text{mR}\,\text{h}^{-1} - 10^4\,\text{R}\,\text{min}^{-1}$
- pulse counting $\quad$ $10^{-6} - 10^{-4}\,\text{R}\,\text{h}^{-1}$
- short-circuit photocurrent $\quad$ $> 10\,\text{mR}\,\text{min}^{-1}$
- change in reverse current $\quad$ $> 1\,\text{mR}\,\text{min}^{-1}$

p–n photoconductive cells are less sensitive than cadmium sulphide crystals, but have an extremely fast response and can be used for energy spectroscopy. Lithium drifted p–i–n is one of the most suitable for gamma-rays. Bulk-induced conductivity in various materials such as diamond, silicon, germanium and GaAs can also be used for X-ray measurement and recording. In a pure material, which is virtually free from electron traps, the conductivity is proportional to the square root of the radiation intensity, and the advantage shared by all solid-state detectors compared with ionization chambers is their small size and the lower intensity required to produce one charge-pair.

To summarize four types of semiconductor X-ray detector can be used [6].

1. Photoresistances with a DC bias (Fig. 5.6(a)): these are not particularly suitable for high-energy radiation because they must be kept physically small to maintain a constant bias and are not therefore efficient absorbers.
2. Converters using p–n and p–i–n junctions (Fig. 5.6(b)): these can be used with or without back bias and the noise is determined by the back current. If the depleted zone occupies almost the whole volume of the detector, the time lag is small.
3. Metal–dielectric–semiconductor–metal systems with a pulsed bias (Fig. 5.6(c)): these are particularly suitable for higher energy radiation and are almost without time lag.
4. Metal–dielectric–semiconductor systems with a high-frequency bias (Fig. 5.6(d)): this circuit arrangement can produce a large amplification of the photocurrent by means of a suitable increase in the time lag.

Matrix detectors

It is obviously possible to construct a matrix of small scintillator crystals or of semiconductor detectors and to use this as a device to record an X-ray image by conversion to light or to electrical signals. A CCTV camera focused on to a fluorescent screen, or a TV camera with its pick-up surface sensitive to X-rays, will do the same. In this recording technique, the image can be digitized in an analogue-to-digital

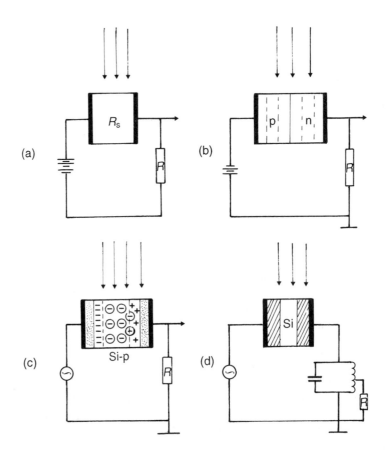

Fig. 5.6 Types of semiconductor detector for ionizing radiation. (a) = photoresistor with resistance and a DC bias; (b) = photodiode using p–n and p–i–n junctions with reverse bias; (c) = metal–dielectric–semiconductor–metal system with pulsed bias; (d) = metal–dielectric–semiconductor system with a high-frequency bias.

converter and held in a computer store, where it is available for image processing.

Small panels of light-sensitive elements on a matrix of 1024 elements are already commercially available (Reticons) and these have been built into image intensifier tubes to produce a digital output (Digicons) [7]. Silicon surface-barrier diodes have also been used in computed tomography applications as multi-channel detectors. The use of a CCTV camera with a directly X-ray sensitive screen is also possible and a few have been constructed and used for specialist applications where the small imaging area is acceptable. (Chapter 11).

5.6 FLUORESCENCE

When an atom absorbs energy an electron may be transferred to an orbit of higher energy and the atom is then said to be excited. The excited state, however, is unstable and the electron eventually returns to its original state, with the emission of the excess energy as electromagnetic radiation. Stokes' law, which applies to reactions of this kind, states that the energy emitted is less than that absorbed. If, therefore, the excited state was induced by the absorption of a quantum of X-radiation, the emitted radiation must be of longer wavelength. This phenomenon is known as 'fluorescence', and has already been mentioned in connection with salt intensifying screens. Substances which exhibit fluorescence with the emission of visible light are known as phosphors, so that if a phosphor is placed in an X-ray field a visible indication of the X-ray image is obtained.

Both inorganic and organic substances which exhibit fluorescence when irradiated with X-rays exist. The inorganic substances usually possess a high light yield; thallium-activated sodium iodide is one of the most widely used materials. Certain plastics can also be used as phosphors and they have an advantage in that they can be more easily prepared in any shape or size.

Normally an electron in an excited state returns to its original level within a fraction of a microsecond and in this case the fluorescence ceases immediately the exciting radiation ceases. There are, however, some substances from which the emission of light persists after excitation. This effect is known as 'phosphorescence', and the emission of light may last for periods extending from a fraction of a second to hours after excitation. The effect, commonly known as 'afterglow', is an undesirable property of certain types of salt intensifying and fluoroscopic screens.

The colour of the fluorescent light is determined by the change in energy as the electron drops to its original level, but most phosphors fluoresce in the yellow–green to blue region. The colour can be markedly influenced by the presence of minute traces of impurities; thus, zinc sulphide, which normally emits blue light, can be made to fluoresce reddish–orange by the addition of manganese, or green by the addition of silver. It is therefore possible, by the addition of minute quantities of metallic impurities called activators (during manufacture) to alter the quality and the quantity of the fluorescence.

Fluorescent screens

If the phosphor is to be used as a fluoroscopic screen the emission should preferably match the sensitivity of the eye under low brightness conditions, when the eye's maximum sensitivity is at a wavelength of about 5500 Å. The screen should therefore have its maximum intensity of emission at about this same wavelength. Zinc cadmium sulphide phosphors satisfy this requirement and this material is now the most commonly used fluoroscopic screen phosphor. Barium platinocyanide, which fluoresces green, has also been used for fluoroscopic screens in the past.

For use as an intensifying screen, a phosphor is usually chosen which has its emission in the blue or ultra-violet region, since photographic film has its maximum sensitivity in this region. Until recently the most commonly-used phosphor was calcium tungstate, but a series of phosphors based on the rare earths, gadolinium, terbium and lanthanum, have been developed, which have increased the effective speeds of the fastest screens used in medical radiography by a factor of ~10. In all these screens the phosphor is in the form of a finely-ground powder coated in a binder on card or plastic as a thin uniform layer. As these new screens have their maximum emission in a different part of the spectrum, special (screen-type) films are needed.

Xeroradiography

A thin deposit of photoconductive material on a backing metal plate, held in the dark, will hold an electrostatic charge until irradiated, when the local conductivity of the layer is increased in proportion to the intensity of the incident radiation falling on different parts of the layer. The electrostatic charge then leaks through to the backing plate and a spatial electrostatic latent image is left on the surface, which can be made visible by dusting with an appropriately charged fine powder; the quantity of powder adhering to the plate locally depends on the residual charge on different areas of the plate. This is the principle of xerography, where the incident radiation is visible light, and is the basis of the well-known Xerox copying systems. With X-rays the same process operates. The photoconductive layer is the same, vitreous amorphous selenium, and the process is called xeroradiography [8]. The process has found only limited use in industrial radiography as the need to use rigid plates, the difficulties of coating these plates with a thick layer of selenium and the limited life of the plates, has restricted its wider application. The typical xeroradiographic image with its enhanced edges is very attractive for certain applications such as the radiography of assemblies, and the equipment can be built as an automatic system in which the plate is exposed, powder-developed, then after examination the plate is cleaned for re-use and is recharged.

Electroluminescent materials

It is possible to construct a panel consisting of a layer of photoconductive material backed with a layer of electroluminescent material which, when an applied voltage is connected across the total thickness, acts as a fluorescent screen that is capable of producing an image several times brighter than a conventional screen. Unfortunately, the light amplification factor is not constant with input intensity. In the combination, the photoconductor acts as a variable resistance controlling the flow of potential into the phosphor. With a sensitive phosphor a low-level X-ray intensity can control a high-output light level. The two layers, the photoconductor and the electroluminescent phosphor, are coated on opposite sides of an opaque insulating layer to prevent light feedback to the photoconductor. Electrodes are provided on the outer surfaces, the electrode against the phosphor being transparent since this is the viewing side, and an AC potential is applied across the combination. The layers are normally very thin and currents excited in the photoconductor by a pattern of X-rays do not spread sideways, so a sharp image is obtained. Typically, the photoconductor can be cadmium sulphide activated with copper and gallium, and the electroluminescent layer is zinc sulphide activated with copper and aluminium [9].

Another method, developed by the company Fuji [10], is the laser-stimulated-luminescence phosphor plate. A support plate is coated with a layer of europium-activated barium fluorohalide, which stores X-ray energy in a quasi-stable state, i.e. as a latent image. Stimulation by laser light from a He–Ne laser causes emission of luminescent radiation corresponding to the intensity of the absorbed X-ray energy. This light is collected by a lens and is directed through a light guide on to a photomultiplier tube. The image is recorded by laser printer, and after use the phosphor plate can be flooded with light to remove any residual image and can be re-used. The light response to X-ray dose is linear over a very large range and the image can be scanned with a $100\,\mu m$ spot on a $10\,\text{pixel}\,mm^{-1}$ raster. So far as can be discovered, these plates have not been used for industrial applications. An alternative scheme, instead of using powder to make the image visible, is to scan the electrostatic image with a small detector (microelectrometer) and feed the data obtained into a storage system [11]. A method based on the same principle of having an electrostatic charge image and a powder developer has been proposed using a sheet of plastic insulator instead of amorphous selenium.

Microchannel plates

These are perhaps more appropriate to Chapter 11 on image intensifier systems, but because of the principles involved they will be mentioned here. A microchannel plate consists of a series of very small cells between two parallel plates across which there is a high voltage (*c.* approximately 1 kV), and for X-ray use one of the plates or an additional plate adjacent to it must be an X-to-electron converter such as a thin gold or tantalum foil. In each cell an incident electron produces an electron avalanche by secondary emission, which is then converted to a visible image by a suitable conversion screen. The output electrons are further accelerated, by a second high voltage, on to the output screen and the total 'gain' in the process can be as high as 10^5. Microchannel plates up to 150 mm diameter have been constructed [12].

REFERENCES

[1] Farnell, G. C. and Broadhead, P. J. (1978) *Photo. Sci.*, **26**, 7.
[2] Bromley, D. and Herz, R. H. (1950) *Proc. Phys. Soc. Lond.*, **63**, 90.
[3] Herz, R. H. (1949) *Photo. J.*, **89B**, 89.
[4] Halmshaw, R. (1971) *Photo. J.*, **19**, 167.
[5] Gifford, D. (1984) *Handbook of Physics for Radiologists & Radiographers*, Wiley.
[6] Gorbunov, V. I. and Melikhov, V. S. (1978) *Defektoscopia*, (2), 28.
[7] Choisser, J. P. (1977) *Opt. Eng.*, **16**(3), 262.
[8] Durant, R. L. (1957) *Further Handbook of Industrial Radiology*, Edward Arnold, London, Chapter 14.
[9] Diemer, G. and Klasens, H. A. (1955) *Philips Res. Rep.*, **10**, 401.
[10] Halmshaw, R. and Ridyard, J. N. A. (1990) *Brit. J. NDT*, **32**(1), 17.
[11] Boag, J. W. (1979) *Phil. Trans. Roy. Soc. Lon.*, **292(A)**, 273.
[12] Chalmeton, V. (1977) *Acta Electronica*, **20**(1), 53.

6

Radiographic techniques: principles

6.1 INTRODUCTION

Ideally, the choice of a technique ought to be made in terms of a required defect sensitivity, but it is seldom possible for users to state categorically the size of the smallest defect which they wish to detect. Consequently, the choice of technique is frequently made in terms of some other factor such as convenience or availability of equipment, time involved etc. Not all radiographs are required to have high defect sensitivity, but it is seldom that sensitivity is not an important factor. It is necessary, therefore, to understand the interplay of the various parameters in each technique from the viewpoint of their effect on sensitivity.

The specimens to which radiography is applied can be broadly divided into three groups:

1. uniform-thickness specimens, which are almost always examined to detect internal flaws;
2. non-uniform thickness specimens, again examined to detect internal flaws;
3. non-uniform thickness specimens, such as assemblies, which are radiographed to check correctness of assembly, presence of components etc., but not usually for the presence of small material flaws.

In all three groups there may be a need for high- and low-sensitivity techniques. In the high sensitivity case, all the technique factors are adjusted to make the defect sensitivity (and the image quality indicator sensitivity) the best possible. An example of a high-sensitivity requirement is the inspection of manual arc-welding of metals; an example of a low-sensitivity requirement would be the inspection of a structural unstressed casting.

6.2 EQUIPMENT DATA

Thickness penetration

The first requirement in radiography is radiation which is capable of penetrating the specimen thickness, with a set-up which will produce the sensitivity required within an economical exposure time. Table 6.1 gives practical values based on present day equipment. The 'high sensitivity' column represents the use of fine-grain film with lead or metal intensifying screens such as would be recommended for the radiography of steel butt-welds in pressure vessels. The 'low sensitivity' column represents the use of faster films, also with lead intensifying screens. This latter technique would be unsuitable for the detection of fine cracks.

Exposure curves

The first requirement is to obtain the constants of the equipment itself. These are kilovoltage, current output in milliamperes, focal spot size and the physical position of the focal spot. It is also desirable to know the inherent filtration, particularly if using low kilovoltage equipment.

It is not usually necessary for a purchaser to check the kilovoltage of a modern X-ray set, and indeed it is quite difficult to do so with much accuracy. It is possible to obtain a calibration value by accurate radiography of a metal step-wedge (section 3.4).

Manufacturers rarely give a certificate of focal spot size, and it is desirable to measure this for each X-ray tube, using the method of

Table 6.1 Radiography of steel specimens (maximum thicknesses which can be examined with typical commercial equipment)

X-rays (kV)	High sensitivity technique, maximum thickness (mm)	Low sensitivity technique, maximum thickness (mm)
100	10	25
150	15	50
200	25	75
300	40	90
400	75	110
1000	125	160
2000	200	250
8000	300	350
Gamma rays		
Ir-192	60	100
Cs-137	100	110
Co-60	125	185

pinhole projection described in the Appendix to this Chapter. European legislation may force manufacturers to provide this information.

Curves of exposure against specimen thickness are required for each of the materials that are to be examined and these must be obtained for each individual X-ray set, in the exposure conditions (film, processing etc.) which are going to be used. Manufacturers often supply a family of curves, and it may only be necessary to make a few spot-checks and apply a correction factor to all the curves.

The usual method of making an exposure chart is to choose an arbitrary value of source-to-film distance, a type of film and to standardize processing; then determine the exposures required in (mA min) to produce a standard film density (say 2.0) through various thicknesses of material using a range of kilovoltages. Step-wedges can be used for these experiments, provided that the steps are not too small in area; a minimum step-size should be five times the step thickness, otherwise the scattered radiation is liable to differ significantly from that under a uniform-thickness plate. The test plate should either be considerably larger than the film, or should be adequately edge-masked.

Typical sets of exposure charts obtained in this manner are shown in Figs 6.1–6.3. Each of these is drawn for an arbitrary source-to-film distance for convenience of calculation, but this distance is not the distance to be used in taking radiographs. Exposures required at any other distance are easily calculated from the inverse square law:

$$\frac{\text{(Exposure required at source-to-film distance } y)}{\text{(Exposure required at source-to-film distance } x)} \, y^2/x^2 \qquad (6.1)$$

Exposures required on different films can be determined from a table of film speeds such as those given in Table 5.2. Thus, the film speed for D.4 film is 100 and for D.7 is 400 with the same processing, so D.4 film will require four times the exposure given to D.7 film for the same film density, other conditions being equal.

For variation in film density, the factors in Table 6.2 are sufficiently accurate to apply to most radiographic films. It is also possible to vary the effective film speed by changing the development time, but with the widespread use of automatic processors this is now seldom done. It

Table 6.2 Exposure factor required for different film densities

Density	Factor
1.5	0.7
2.0	1.0
2.5	1.3
3.0	1.6

should be remembered that the film speed is a function of both the film type and the developer and the development conditions used.

The exposures given in Figs 6.1–6.3 are given in (mA min) for X-rays and in (Ci min) for gamma-rays, and with all metal screen exposures the evaluation of exposure time is simple. Thus, 1 min at 10 mA is exactly equivalent to 30 s at 20 mA or 2 min at 5 mA etc. Exactly the same applies to gamma-ray exposures.

Radiography of materials other than steel requires the use of a conversion factor which is dependent on the X-ray energy. Table 6.3 gives approximate practical conversion factors for some materials compared with steel. Equivalent thickness factors in Table 6.3 are to be read off vertically only. These equivalent thicknesses are based on typical thicknesses which would be radiographed with each radiation energy. For example, with 100 kV X-rays the steel/aluminium ratio is 1:13 so an exposure suitable for 80 mm aluminium would be suitable for 6 mm steel.

Exposure data such as are given in graphical form in Figs 6.1–6.3 can be set out in the form of a nomogram, or built into a slide rule, or developed as part of a computer program.

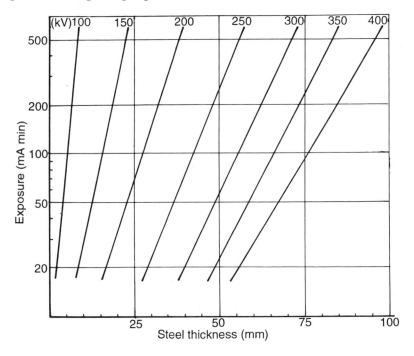

Fig. 6.1 Exposure curves for steel plates and X-rays between 100 and 400 kV; D.7 film; 1000 mm source-to-film distance; film density 2.0; lead-foil intensifying screens.

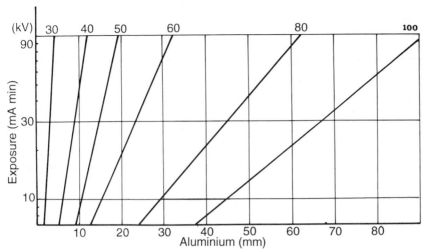

Fig. 6.2 Exposure curves for aluminium and X-rays from 30 to 100 kV; D.7 film; density 2.0; no screens; 900 mm source-to-film distance.

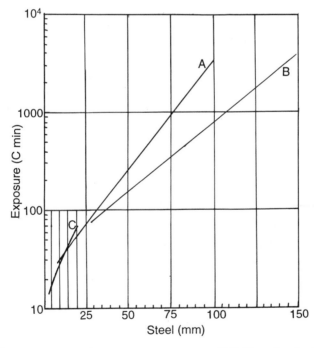

Fig. 6.3 Exposure curves for gamma-ray sources; D.7 film; density 2.0; 5 min development in Agfa-Gevaert VC developer at 20°C. Curve A = iridium-192 source; lead screens; 500 mm source-to-film distance; Curve B = cobalt-60 source; copper screens; 500 mm source-to-film distance; Curve C = ytterbium-169 source; lead screens; 200 mm source-to-film distance.

Table 6.3 Approximate equivalent thickness factors

Metal	X-rays (kV)							Gamma-rays	
	50	100	150	200	250	300	400	Ir-192	Co-60
Magnesium	29	20	18	14	–	–	–	–	–
Aluminium	17.5	13	8	6.3	5.5	5.2	4.5	3	–
Dural; 4% Cu	13	11	7.5	5.5	4.8	4.7	–	3	–
Steel	1.0	1.0	1.0	1.0	1.0	1.0	1.0	1.0	1.0
Copper	0.75	0.68	0.7	0.73	0.68	0.65	0.6	0.9	0.9
Brass	–	–	0.8	0.8	0.8	0.8	–	0.9	0.9
Lead	–	–	0.09	0.075	0.065	0.06	–	0.25	0.65

6.3 IMAGE PARAMETERS

As stated, the fundamentals of image quality on films are contrast, definition (unsharpness) and graininess. Contrast depends on X-ray energy, scatter conditions, on the film used, its processing and its density. Graininess is obviously dependent on the type of film and its processing. Definition depends on the geometric conditions of the set-up and on the film characteristics.

   In the final analysis, all three parameters have some dependence on the film viewing conditions and the performance of the eye of the film reader. Here these parameters, although not always independent, will be treated separately using what may be called the 'classical' treatment. Chapter 8 describes a more advanced treatment based on modern concepts of treating an image as a spatial frequency function.

Image unsharpness

In practical industrial radiography there are four possible causes of radiographic unsharpness, neglecting such adventitious effects as bad screen–film contact:

1. geometric unsharpness U_g arising from the finite size of the source of radiation, which is never a true point source;
2. film unsharpness U_f, which is due primarily to scattering of electrons in the film emulsions;
3. Screen unsharpness U_s, which is due to scattering of light in the fluorescent layer of salt intensifying screens;
4. movement unsharpness U_m, which is due to relative movement of the specimen and film during the exposure.

The total effective unsharpness on the film must therefore be some function of one or more separate causes, and the method of summing unsharpnesses must be considered.

Geometric unsharpness

Figure 6.4 illustrates the essential points concerning the magnitude of U_g. By simple geometry

$$U_g = \frac{sb}{a} \tag{6.2}$$

where s is the width of the radiation source as seen from the film, b is the distance from the object detail to the film, and a is the distance from the object detail to the radiation source.

The geometric unsharpness usually calculated is the maximum value, corresponding to an object detail on the surface of the specimen remote from the film; in practice, for real defects inside the specimen, U_g will vary from near zero up to this maximum value, depending on the position through the thickness of the specimen of the object detail being imaged.

If the focal spot of the X-ray tube emits radiation uniformly over its area, the density distribution curve across the image of an object has three straight-line portions and is of the form shown in Fig. 6.4. In

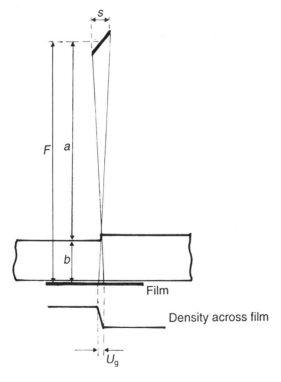

Fig. 6.4 Geometric unsharpness U_g.

Table 6.4 Typical values of geometric unsharpness for a range of industrial radiographic techniques

Radiation	Source widths (mm)	Specimen thickness (mm)	Source-to-film distance (mm)	U_g (mm)
100 kV X-rays	2	6	500	0.02
200 kV X-rays	5	25	1000	0.13
400 kV X-rays	7	75	1000	0.57
Co-60 gamma-rays	4	50	500	0.44
Ir-192 gamma-rays	2	25	500	0.11

practice, the edge of the object must have a finite thickness in order to absorb radiation, and edge effects cause the density distribution curve to have a small toe and shoulder. Also, focal spots of X-ray tubes rarely produce uniform emission of radiation across their surface, and this also leads to a more complex form of density distribution curve with a toe and shoulder. Even this treatment is too simple: most X-ray tube focal spots are markedly non-uniform, emitting perhaps 30% more radiation in bands near the edges than on the centreline. The magnitude of geometric unsharpness in industrial radiography can vary markedly. The majority of industrial X-ray tubes have focal spots of 2 to 5 mm effective diameter, while radioisotope gamma-ray sources are typically 1, 2 or 4 mm in diameter. Using these data, Table 6.4 shows some typical values of U_g.

Before considering other causes of unsharpness, the importance of the problem is well illustrated by consideration of the density distribution across the image of a small defect in a specimen, as the geometric unsharpness is progressively increased. Taking first the case where the radiation source is smaller than the width of the defect, Fig. 6.5(a) shows the idealized case where the focal spot is a true point source. Figure 6.5(b) shows the density distribution in the image when the radiation source is larger. The image now consists of two blurred edges and a central region. If the radiation source is made still larger, the width of the blurred regions increases until they overlap and an image of reduced contrast C, compared with the original image contrast C_0, is obtained (Fig. 6.5(c)). The contrast in this image can be expressed by the ratio

$$C = C_0 \frac{x}{U_g} \left(\frac{a+b}{a} \right) \tag{6.3 a}$$

When $a \gg b$, which is the usual case, this becomes

$$C = C_0 \frac{x}{U_g} \tag{6.3 b}$$

The critical case when the contrast begins to be reduced is when $x = U_g$.

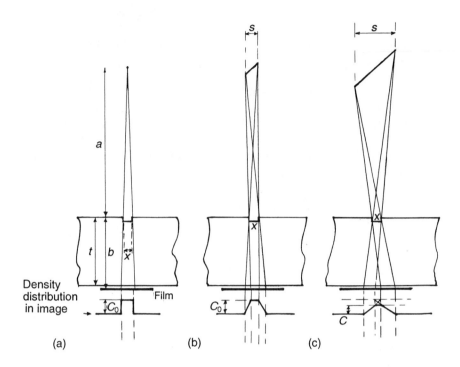

Fig. 6.5 Geometric unsharpness as it affects the image of a small defect. (a) = point source, $U_g = 0$, image contrast is C_0; (b) = larger source diameter; (c) = large diameter source, image contrast is C.

Thus, the effect of geometric unsharpness on the image of a small defect is:

1. to increase the width of the image on the film, compared with the real width of the defect;
2. to blur the edges of the image of a sharp detail;
3. to reduce image contrast when the defect is smaller in width than the source of radiation.

Although these effects have been demonstrated for geometric unsharpness, it is obvious that they are also produced by any other cause of radiographic unsharpness, i.e. the image is broadened, blurred, and in some cases there is a reduction in contrast.

Film unsharpness

Film unsharpness (also sometimes called inherent unsharpness) arises from the generation of secondary electrons in the film emulsion. When a quantum of ionizing radiation is absorbed in a silver halide crystal in the film emulsion there is sometimes sufficient energy both to make the crystal developable and to release secondary electrons with sufficient energy to travel through the emulsion to other silver halide grains, and in turn make these developable. Thus, instead of a single 'grain' from each X-ray quantum absorbed, there is a small volume or string of grains, and from the imaging viewpoint this is equivalent to a reduction in image sharpness because a point is imaged as a small disc. With high-energy radiation, several hundred silver halide crystals may be affected by the absorption of one X-ray quantum. The practical effect is to produce an unsharp image of a sharp specimen detail, with density distributions analogous to those shown for geometric unsharpness. The width of the band of blurring can be taken as a measure of the film unsharpness, in the same manner as for U_g.

There has been a considerable number of papers suggesting precisely what distance on the unsharpness curve should be taken as a measure of U_f, but in practice the alternative methods do not lead to very different numerical results [1]. With an unsharpness curve having a toe and shoulder, such as in Fig. 5.4(c) and (d), the commonest method of obtaining a value of unsharpness is to cut off a portion of the toe and shoulder as shown in Fig. 6.6, as first proposed by Klasens [2]. Other more complex methods have been proposed such as that by Möller and Weeber [3] and by Halmshaw [1].

Salt intensifying screens cause a large unsharpness, due to light scatter in the fluorescent layer of the screen, and usually a separate name is used; screen unsharpness U_s.

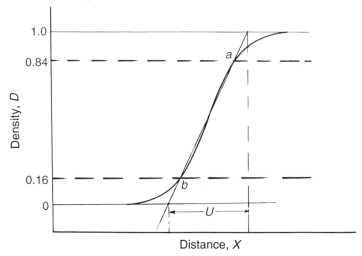

Fig. 6.6 Unsharpness curve, density D plotted against distance x. Klasens' criterion for the value of U. Involves joining points a and b and projecting the line on to the axes at D_{max} and D_{min}.

Metal intensifying screens cause only a very small increase in film unsharpness unless thick screens are used.

The difference in film unsharpness for different radiographic films used at the same radiation energy is very small and there have been no experimental results showing a significant change between any conventional films, when used at ordinary X-ray energies, except for some ultra-fast films which are no longer marketed.

The values of film unsharpness to be used in calculations of source-to-film distances were given in Table 5.3 and are based on the work of several investigators [1–5]. These values refer to filtered radiation such as is obtained after the radiation beam has passed through a specimen.

Combined effect of several causes of unsharpness

Obviously, on most radiographs concern is with the combined effect of at least two causes of unsharpness, geometric and film. One cause may be predominant but both will have some effect on the image. There have been a variety of proposals to calculate the combined effect of two causes of radiographic unsharpness; it was realized at an early stage that unsharpnesses are not directly additive. Berthold [6], the first worker in this field, suggested that if one unsharpness was appreciably larger than the other, the smaller one could be neglected and the total unsharpness should be taken as equal to the largest single unsharpness. Newell [7] proposed

$$U_{\text{total}} = (U_1^2 + U_2^2)^{1/2} \qquad (6.4)$$

and Klasens [2] that

$$U_{\text{total}} = (U_1^3 + U_2^3)^{1/3} \qquad (6.5)$$

U_1 and U_2 in each case being the individual unsharpnesses.

Halmshaw [1] proposed that when the unsharpness curve is of complex shape, a curve of the simple linear form of Fig. 5.4(b) can be drawn to have the same maximum and minimum densities and the same area under the curve. Using this equivalent area assumption, both the shape and width of the curve for a combination of unsharpnesses can be calculated. If the two individual unsharpnesses can be regarded as linear then

$$U_{\text{total}} = U_2 + U_1^2/3U_2 \qquad (6.6)$$

and if one unsharpness (U_f) is regarded as an exponential curve, the other being linear, then

$$U_{\text{total}} = U_g + U_f^2/2U_g - U_f^2/2U_g \left[\exp\left(-2U_g/U_f\right)\right] \qquad (6.7)$$

Equations 6.4–6.7 do not give very different results numerically when typical values are inserted, and the differences between them, except in abnormal examples, are probably too small to be resolved by direct experimental checks. The simple sum-of-squares formula (equation 6.4) can therefore be used for most practical cases.

With typical industrial radiographic films, the curve of film density against radiation intensity is approximately straight at the film densities employed, and when, as when considering unsharpness curves, there is only a very small density range, film density and radiation intensity can be interchanged without any correction for nonlinearity.

Any radiographic image of a sharp-edged detail is therefore blurred. Knowing the causes of this blurring, the magnitude of the effect can be calculated for a range of conditions typical of industrial radiography. In going from, say, 10 mm aluminium radiographed with 50 kV X-rays to 150 mm steel radiographed with cobalt-60 gamma-rays, it is found that the total unsharpness may vary from 0.02 mm to 1.5 mm, a range of about 70:1. Also, as an example, in radiographing 25 mm steel welding with 200 kV X-rays to search for cracks of the order of 0.025 mm wide or less, the acting unsharpness may then be many times larger than the width of these cracks. These examples show that radiographic unsharpness is usually of such a magnitude as to have an important influence on the visibility of detail on a radiographic image.

It is appropriate, next, to consider the effect of unsharpness on the density distribution in the image of different types of flaws. A brief outline of some of these effects was given when illustrating the result of increasing the geometric unsharpness (Fig. 6.5), and exactly the same conclusions are valid for the effect of any radiographic unsharpness or combination of unsharpnesses on any image.

1. The width of the image obtained w is the width of the object detail x plus the total unsharpness U:

$$w = x + U$$

2. Equation 6.3 can be re-expressed in a more general form as

$$C = C_0 \frac{x}{U_t}, \quad U_t > x \tag{6.8}$$

where U_t is the total unsharpness.

Examples of these effects are illustrated in Fig. 6.7. The density distribution across the image of a narrow idealized slit or crack is as shown when there is no unsharpness; the effect of a total unsharpness less than the slit width is to broaden the image, but not alter the contrast; as soon as the total unsharpness is made greater than x, however, a reduction in contrast commences and a broad low-contrast image results.

To give a practical example, if a crack of width 0.025 mm is radiographed under conditions such that the total unsharpness is 0.2 mm, the

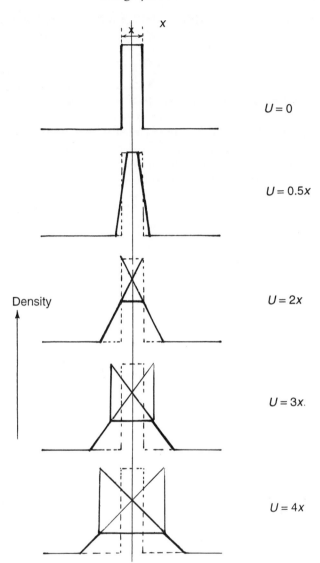

Fig. 6.7 Density distribution curves showing the effect of a simple linear unsharpness on the image of an idealized slit (width x) for different values of U/x.

contrast is reduced by a factor of eight from the theoretical maximum value. In Fig. 6.7, for simplicity, a simple linear unsharpness has been taken; if the unsharpness curve is assumed to have a toe and shoulder, similar curves can be constructed. It will be noted, therefore, that the images of narrow cracks in line with the X-ray beam are broadened into lines approximately the width of the acting unsharpness when this is considerably greater than the crack width, so that from the view point of interpreting the radiograph the true width of a very narrow crack cannot be determined.

A further example is also worth considering. The thickness variation across the cross-section of a circular section wire can easily be calculated from the basic X-ray attenuation law. Using the film density relationship ($D \propto \log_{10} I$) which is valid for small density differences at the densities normally used in radiography, this can be taken as the curve of density distribution across the image of the wire, assuming no unsharpness. The influence of unsharpnesses of different magnitudes on the shape of this curve can now be calculated either by summation or by other methods. The results are shown in Fig. 6.8. It will be seen that in the case where the

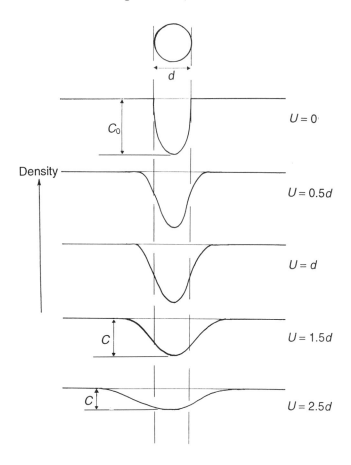

Fig. 6.8 Density distribution curves showing the effect of unsharpness on the image of a wire for different values of U/d.

unsharpness is equal to the wire diameter, the image of the wire is considerably distorted, the rate of change of density across the image being much reduced, but the contrast is unchanged. If the unsharpness is made proportionally larger, similar effects to those shown in Fig. 6.7 are produced. The image becomes still more broadened and reduced in contrast.

These concepts of unsharpness, film and geometric, lead a considerable way towards understanding the fundamentals of the radiographic image on film. They enable the degradation of the image from the ideal to be explained and predicted, and curves of the density distribution in the image of a specimen detail can be derived. They also enable one of the fundamental parameters of a radiographic technique, the source-to-film distance, to be calculated.

Source-to-film distance

Obviously, to obtain a high quality radiograph capable of detecting fine detail, the value of total unsharpness must be kept to a minimum, which implies that the individual unsharpnesses should be minimized. This, however, is not possible without compromises. If U_g is to be reduced for a given X-ray source size (effective focal spot size) and for a given specimen thickness, this can only be done by increasing the focus-to-film distance, which will require longer exposure times according to the inverse square law. Experimental studies using small linear defects such as narrow cracks have shown that the sensitivity improves as the source-to-film distance is increased, when all other parameters except exposure times are kept constant (i.e. the film density, and hence the film gradient, is constant).

Two typical experimental results from a large series are shown in Fig. 6.9 (curves A and B). These show that there is an increase in crack sensitivity until at least $U_g = U_f$. However, similar curves for wire and small holes (as in IQIs, Chapter 7) show much less change (Fig. 6.9, curves C and D). In Fig. 6.9 (curve C) the wire sensitivity hardly changes between $U_g = 2U_f$ and $U_g = 0.5U_f$. A probable explanation of this result is that the image of a wire is still discernible even when it is considerably blurred, whereas the image of a finer crack disappears much more rapidly due to loss of contrast.

The choice of appropriate values of source-to-film distance is one of the most contentious topics in determining a satisfactory radiographic technique, and existing standards from various countries differ widely. Obviously the major parameter is U_g, but there is no point in reducing U_g if U_f is large (as when using gamma-rays or high energy X-rays). The acceptable maximum value of U_g should therefore be a function of the radiation energy to be used, and therefore of U_f. From Fig. 6.9 (curve A),

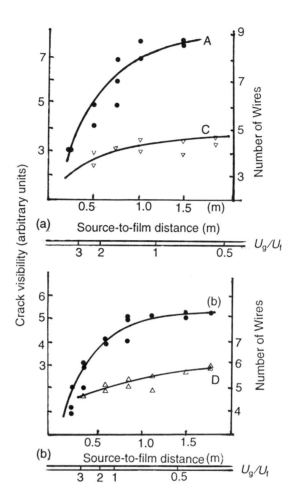

Fig. 6.9 The effect of source-to-film distance on the discernibility of cracks and wires. (a) = 25 mm steel specimen, X-rays, MX film; (b) = 75 mm steel specimen, Co-60 gamma-rays, D.7 film. Curves A and B are for cracks; curves C and D for wires. The experimental points shown are the results of several radiographs and several readers.

for a high-sensitivity technique designed to detect narrow cracks it would be desirable to make $U_g = U_f$, but this may sometimes lead to uneconomically long exposure times, although often these longer exposures can be reduced by using a slightly faster film or a slightly higher kilovoltage. It is at this stage that the specification of a radiographic technique ceases to be an exact science and where experience and a certain amount of art is involved in striking a compromise between the

Table 6.5 Minimum source-to-film distances: practical recommendations for a high-sensitivity technique in (steel)

Steel Specimen thickness (mm)	Approximate radiation	Source-to-film distance (mm)	
		4 mm *diameter* source	2 mm *diameter source*
12	150 kV X-rays	540	280
25	220 kV X-rays	820	420
37	270 kV X-rays	1110	560
50	300 kV X-rays	1370	710
63	350 kV X-rays	1610	820
75	400 kV X-rays	1600	–
100	400 kV X-rays	2010	–
25	Ir-192 gamma-rays	610	320
50	Ir-192 gamma-rays	1220	640
75	Co-60 gamma-rays	750	400

various factors. For example, a reduction of 20% in the source-to-film distance from the condition $U_g = U_f$ will cause only a small deterioration in sensitivity, while allowing a 36% reduction in exposure time. This reduction of 20% would lead to the source-to-film distances shown in Table 6.5. Apart from the 100 mm thick steel specimen, these values of source-to-film distance and kilovoltages would result in exposure times which are not impractically long, and it is suggested that these values should be used for high-sensitivity radiography, with the proviso that a slightly better crack sensitivity can be obtained by increasing the distances.

Proposals for minimum source-to-film distance in various national and international standards [8–10] can be summarized as follows.

1. ISO: 5579, EN-444 and DIN-54111 all use an unexplained formula to determine the minimum source-to-film distance:

$$a/s = K\, b^{2/3} \tag{6.9}$$

   where a is the source-to-object distance (in mm), s is the effective source diameter (in mm), b is the specimen thickness (in mm), and K (a constant) = 15 for high-sensitivity techniques (called class B), and $K = 7.5$ for 'normal' techniques (class A). The source-to-film distance is $(a + b)$. Class A is considered to be 'standard' radiography. This formula gives a complex relationship between maximum acceptable values of U_g and radiation energy: over the middle range of thicknesses K = 15 gives values of source-to-film distance similar to those obtained by taking $U_g = 2U_f$, but for thick sections and for K = 7.5, very much smaller source-to-film distances are permitted.

2. BS: 2600:1983 and BS:2910:1986 give graphs of minimum acceptable values of source-to-film distance for two techniques called 'normal'

Table 6.6 ASTM values for geometric unsharpness

Specimen thickness (mm)	*Maximum value of* U_g
Under 51	0.02 in (0.51 mm)
51–77	0.03 in (0.76 mm)
57–102	0.04 in (1.02 mm)
Greater than 102	0.07 in (1.78 mm)

and 'more critical', depending on whether medium-speed or fine-grain film is employed, respectively. The source-to-film distances for the 'more critical' technique correspond broadly to $U_g = 2U_f$ and for the 'normal' technique to about 70% of these values.

3. ASTM.E.1032 (for weldments) specifies maximum values of geometric unsharpness, as shown in Table 6.6. These are described as 'for all applications, unless otherwise specified'. The justification for such large values is not explained, and they appear to sacrifice good flaw sensitivity to enable shorter exposure times to be used.

Li Yan [11] reviewed international standards on the subject of source-to-film distances and reached similar conclusions, in that for the detection of small, narrow cracks, the technique parameters should be adjusted to give $U_g = U_f$ or even $U_g = 0.5\,U_f$. He also considered the importance of maximum kilovoltage and minimum density.

Some megavoltage equipments, notably some betatrons, have a very small focal spot size and, because of their high energy, a large film unsharpness, so it ought to be possible to use very small values of source-to-film distance. However, the beam width of the X-ray field is usually small and suitable source-to-film distances are chosen with regard to the size of film required, rather than considerations of U_g and U_f. With many megavoltage equipments such as large linacs, the X-ray output is so large that very large source-to-film distances (e.g. 2 m) can be used without the need for long exposure times. The correct source-to-film distances for light-alloy radiography will differ from the values proposed for steel specimens because, for a given specimen thickness, lower energies will be used, corresponding to smaller values of U_f, so larger source-to-film distances are therefore justified.

6.4 CHOICE OF RADIATION ENERGY

This parameter has already been discussed to a limited extent. Table 6.1 gave a guide to the maximum thickness of steel which can be examined with different radiations. Later, in choosing the appropriate value of film unsharpness, it is necessary to know the approximate kilovoltage for each thickness (Table 6.5).

The importance of choosing the correct kilovoltage varies considerably with the kilovoltage range being considered. For X-rays below 150 kV the choice of correct kilovoltage is critical because the attenuation coefficient varies rapidly. From 200 to 400 kV, only a considerable difference in kilovoltage, of the order of 30–40 kV, will make a significant difference in sensitivity, and in the high-energy region, kilovoltage is relatively unimportant in terms of attainable sensitivity; radiographs taken with 2, 5, 18 and 31 MV X-rays will produce similar flaw sensitivity values.

A good rule-of-thumb when using X-ray kilovoltages below 400 kV is to keep the exposure within the range 10–30 mA min. Usually the only effect of using too low a kilovoltage on a uniform-thickness specimen is that the exposure time will become impractically long; if a long exposure time is tolerable, the flaw sensitivity will be slightly better. The effect of using too high a kilovoltage is to shorten the exposure time required, to lose image contrast and so lose flaw sensitivity. As guide values, to a halve the exposure time requires an increase of 25 kV at 200 kV, or 50 kV at 400 kV. These changes in kilovoltage will produce a loss in sensitivity of about 25%, so that in this region other factors such as film density and source-to-film distance tend to be more significant than kilovoltage.

Gamma-rays

Since it is not possible to vary the energy of radiation emitted by a gamma-ray source, it is necessary to state a thickness range which may be satisfactorily examined with each specific radioisotope. The upper limit is controlled by the commercially-available source strengths, and the lower limit by the fall-off in flaw sensitivity which is inevitable when a fixed energy of radiation is used on progressively thinner specimens. The upper limits are relatively easy to calculate and are not very controversial, but the lower limits are less easy to agree because of the desire to use gamma-rays for many applications for reasons of convenience and portability of the source (Table 6.7). If gamma-rays are used on smaller thicknesses than given on Table 6.7 for reasons of convenience, it must be clearly understood that these are not good techniques, and there will be a poor sensitivity. None of the first four gamma-ray sources in

Table 6.7 Steel thickness ranges for which gamma-ray sources are suitable

Source	High-sensitivity technique (mm)	Low-sensitivity technique (mm)
Ir-192	18–80	6–100
Cs-134	25–100	20–120
Cs-137	30–100	20–120
Co-60	50–150	30–200
Yb-169	1–5	1–10

Table 6.7 are suitable for light-alloy inspection except on large thicknesses. Experience with Yb-169 on light-alloys is very limited, but good radiographs appear to be possible.

6.5 FILM AND INTENSIFYING SCREENS

There are so many parameters involved in a radiographic technique that it is desirable, at the outset to make some simplifying assumptions in respect of films. The following assumptions are proposed:

1. neither very high-speed film (film speeds of 500 or greater (Table 5.2)) nor salt intensifying screens will be used;
2. normal techniques will make use of films with speeds in the range 50–300, with a preference for the so-called fine-grain films (speeds 50–100) if narrow flaws such as small cracks are likely to be present in the specimen;
3. normal processing techniques will be used, i.e. neither shortened nor extended development times;
4. the radiographs will be exposed to obtain a minimum film density of approximately 2.0 through the image of the most important part of the specimen.

As was shown in Fig. 5.2, for fine-grain films, and to a slightly lesser extent medium-speed films, film contrast increases with film density and continues to increase with higher densities, even beyond those which can be used. When maximum contrast is required, one should therefore, in theory, expose the radiographs to obtain these very high densities after processing. However, one has to 'read' the films, so it is no use having a radiograph of such a high density that one cannot see through it adequately when it is placed on an illuminated viewing screen. The ability of the human eye to discern detail in an image varies markedly with the luminance, and deteriorates rapidly below 10 cd m^{-2}. In practice this means that the maximum film density cannot be higher than 2.5–3 unless special high-intensity illuminators are available for film viewing. The specification of working film density, assuming it is obtained by good film processing, is by far the most important single factor in radiographic techniques. Taking a typical radiographic film, Table 6.8 shows the percentage loss in contrast if the working density is reduced.

Table 6.8 Contrast loss with film density

Film density	Contrast as a percentage of the contrast at density 2.0
3.0	141
2.0	100
1.5	76
1.0	49

Although not of great importance to industrial radiography, it should be noted that with salt screen-type films used with salt screens the film contrast reaches a maximum value at a film density of about 1.3, and for higher densities the contrast is markedly lower. This density is therefore the correct working density for salt screen exposures. To industrial radiographers, who are used to non-screen film exposures, such radiographs appear to be thin and under-exposed.

Film graininess

With low energy X-rays it is good practice to use fine-grain film or, in very critical applications, very fine-grain film. However, the use of the latter, because it is very slow and so requires a larger dose of radiation to achieve an adequate film density, sometimes involves unacceptably long exposure times.

With high energy X-rays, fine-grain and very fine-grain films can be used with most equipment without any problems of long exposure times.

With gamma-rays, fine-grain film can usually be used but very fine-grain films are generally too slow.

At the other end of the film-speed scale, high-speed films are so grainy that for most codes of good practice they are not recommended for high quality radiography.

The concepts of film graininess are quite difficult to handle on a quantitative basis, particularly on films exposed to X-rays. Graininess cannot be controlled to any extent, except by choice of film type, so it will not be discussed further in this section, but it is of great importance in determining the limits of discernible image detail, and is discussed further in Chapter 8.

Intensifying screens

The type of film proposed for all techniques is a non-screen 'direct' type and will therefore generally be used with metal intensifying screens. The thickness of these screens is not generally critical so far as radiographic sensitivity is concerned, but to maintain short exposure times some limits should be imposed. The recommended screen thicknesses are given in Table 6.9.

Polycrystalline (salt) screens are nowadays rarely used in industrial radiography except for a very few special applications such as flash radiography.

There are cases where it is desirable to use an extra thickness of metal screen between the specimen and the film, as a filter, and it is convenient in practice to simply use a thicker front screen. This will be discussed under 'filtration' as a separate parameter in technique, although in one sense all front intensifying screens are also filters.

Table 6.9 Metal intensifying screens

Radiation	Screen material	Front screen thickness (mm)	Back screen thickness (mm)
120 kV X-rays	Lead	None*	0.1 minimum
120–250 kV X-rays	Lead	0.025–0.05[†]	0.1 minimum
250–400 kV X-rays	Lead	0.05–0.15	0.1 minimum
1 MV X-rays	Lead	1.5–2.0	1.0 minimum
5–10 MV X-rays	Copper	1.5–2.0	1.5–2.0 minimum
15–31 MV X-rays	Tantalum[‡]	1.0–1.5	None
Ir-192 gamma-rays	Lead	0.05–0.15	0.15 minimum
Cs-137 gamma-rays	Lead	0.05–0.15	0.15 minimum
Co-60 gamma-rays	Copper Steel[§]	0.5–2.0	0.25–1.0 minimum

*A thin front screen (0.02–0.04 mm) is often used with X-rays between about 80–120 kV to absorb some of the soft scattered radiation generated in the specimen; it does not produce any intensifying action, and functions as a filter.
[†]Thinner lead will produce more intensification, but is not generally available except in integral screens.
[‡]Tungsten screens of the same thickness may also be used.
[§]Copper or steel screens will produce a better radiograph than lead screens, but require double the exposure time for lead screens.

6.6 FILTRATION

The use of a filter close to the X-ray tube to produce a 'harder' beam of radiation and so extend the thickness latitude is one of the simpler aspects of filtration techniques. Typical thicknesses of filter are

- with 200 kV X-rays: 0.25–0.5 mm lead
- with 400 kV X-rays: 0.6–1.0 mm lead

and these are usually placed close to the X-ray tube window.

   One of the most important problems in radiography is the case where the edge of the specimen, or part of the edge, is within the film area and it is required to obtain good flaw sensitivity up to the edge. An associated problem which also includes a thickness latitude component is the radiography of a solid cylinder with the radiation directed along a diameter. In both cases, edge masking can be employed (see below) but this is often a laborious procedure and generally filtration techniques can be used to produce a considerable improvement.

   If the filter is placed close to the X-ray tube it absorbs the very soft radiation emitted by the tube, and this is the radiation most easily absorbed by the film and which has the greatest photographic effect. Without a filter this radiation only reaches the film where there is no specimen, or through the thinnest parts of the specimen, and in these regions it produces an intense film blackening. Thus, in the case of the radiography of a cylinder, there is 'undercutting' of the edges of the image and detail is obscured. Another way of regarding such a filter is to consider it as adding an increment of thickness over the whole specimen. The use of a filter on the tube seems to have most value with X-rays between 150 kV and 400 kV.

The radiation reaching the film is a mixture of the higher energy components of the primary beam and scattered radiation generated in the specimen, and because of Compton scatter much of this scattered radiation is of lower energy. A thin lead filter placed under the specimen will absorb a larger proportion of this lower energy radiation without itself generating as much additional radiation, and so the ratio of scattered-to-direct radiation reaching the film will be reduced. If, therefore, a specimen generates a large amount of scattered radiation or has portions which are not close to the film, a filter between the specimen and film is likely to give the best results. Such a filter may take the form of a thick front intensifying screen [12].

If an increase in latitude is required, or if the edge of a specimen is within the film area, it may be advantageous to put the filter close to the X-ray tube. In either case, typical filter thicknesses are

- 150 kV X-rays: 0.25 mm lead
- 200–250 kV X-rays: 0.5 mm lead
- 400 kV X-rays: 0.6–1.0 mm lead
- 1000 kV X-rays: 1.0–1.5 mm lead.

For uniform-thickness specimens it has been found [13] that a filter of (0.5 mm lead and 0.5 mm tin), placed between the specimen and the cassette, will give a considerable improvement in contrast on specimens thicker than 40 mm steel, using 200–400 kV X-rays. Thicker filters give a little improvement but, as they require longer exposure times, are uneconomic. On thinner sections, although such a filter reduces the build-up factor, it also has the effect of 'hardening' the radiation, so slightly reducing the linear attenuation coefficient μ and increasing the unsharpness. The overall effect, then, is said to be a negligible improvement in image quality.

Claims have been made at various times that an improvement in image quality can be obtained on uniform-thickness specimens radiographed with high energy X-rays or gamma-rays, by having a lead or tungsten filter between the specimen and film of thickness of the order of 1–2 mm. Little experimental evidence has been traced to support this claim.

6.7 MASKING

If the edge of a thick specimen is within the film area it is difficult to eliminate all 'undercutting' of the image, even with filtration techniques, and the edges must be masked to prevent unwanted radiation reaching the film. There are many possible methods, varying in their efficiency and the laboriousness of their application.

1. Lead sheet. Sheets of lead of suitable thickness to produce similar absorption to the specimen are cut to fit around the edges. As a rough guide, 3 mm lead is needed on a 25 mm steel thickness, and 8 mm lead on a 50 mm specimen, and the lead must fit well with no gaps. This is therefore a practical method for straight edges, but not for complex shapes unless the mask can be used repeatedly.
2. Lead salt solution. The specimen is put into a plastic tray and a strong solution of lead salts is poured around it. A suitable solution for steel specimens is 500 g of lead acetate in 1 litre of hot water, to which 400 g of lead nitrate is later added. This solution is only suitable for X-rays below 200 kV as it is not possible to obtain adequate densities of solutions suitable for higher energies. The method has several serious disadvantages: the lead solution is poisonous and corrosive, and in practice the liquids tend to get under the specimen and produce spurious patches on the image. Carbon tetrachloride can be used for masking light-alloy specimens.
3. Lead paste. A paste can be made of Plasticine and lead powder or litharge or of lead salts with grease, and this can be pressed around irregular edges. It can be used to supplement lead sheet. This is a good practical method, and if the paste is correctly compounded it can be relatively clean to use.
4. Lead shot. Fine lead shot, 0.25 mm diameter, can be used. This material 'flows' without running badly, but must be prevented from getting under the specimen. This is another good practical method. Tungsten powder is also used for the same purpose.

When a large number of specimens of the same shape are to be examined, the use of specially-made thickness-equalizing pieces is worth considering; these are shaped to compensate for the thickness differences.

6.8 SCATTERED RADIATION

Scattered radiation reaching the film is one of the most potent causes of reduction in image quality, particularly with X-rays between 150 and 400 kV. The masking and filtration techniques already described can reduce the effects of scattered radiation generated inside the specimen, but scattered radiation can originate from outside the specimen and reach the film:

1. radiation scattered back on to the film after having passed through it, either by the cassette support or the walls or floor behind or below;
2. radiation scattered from primary radiation striking floors or walls (or other large objects in the exposure room);

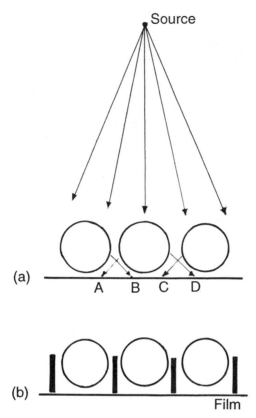

Fig. 6.10 (a) Radiography of a series of cylindrical specimens. Scatter bands are produced at A, B, C and D unless the masking system shown in (b) is used.

3. scatter from one specimen on to the image of a second specimen or another part of the same specimen.

A typical example of the third effect is shown in Fig. 6.10. Sharp, narrow bands of higher density will be seen on the radiograph at A, B, C and D unless lead sheets are placed between the cylinders as shown in (b). It is good radiographic practice to mask the radiation beam to the minimum required area, whenever possible, and to back the cassette with a lead sheet at least 3 mm thick whenever scattering objects behind the film cannot be eliminated. Modern X-ray cassettes often have such lead backing built in.

In general, if a beam restricter cannot be used, such as when panoramic exposures are made, the exposures should be made in as large a room as possible, so that extraneous scatter is attenuated by distance. The specimens, whenever possible, should be well above floor level, and the floor near the specimens should be covered with lead sheet.

Anti-scatter grids

In medical radiography, devices known as fixed grids or moving grids are used to reduce the amount of scattered radiation reaching the film. Such grids are rarely used in industrial radiography, possibly because no satisfactory design has been developed. Another important method of minimizing scattered radiation is to space the film cassette away from the specimen. For X-ray tubes with conventional focal spot sizes (2–5 mm) this would obviously lead to a serious loss of image sharpness, but with microfocus tubes it has become an important technique and is described in more detail in Chapter 7. Some recent work on the measurement of scattered radiation (references [8–12] in Chapter 2) has suggested that with a uniform-thickness specimen, masking down the area exposed has little effect in reducing scatter unless the beam width is less than 20 mm, but an increase in film-to-specimen distance (say, 20 mm stand-off for a 20 mm thick steel specimen) can reduce the build-up factor from 2.6 to 0.6.

6.9 SUMMARY OF PROCEDURE FOR SPECIFYING A TECHNIQUE

For a uniform-thickness or near uniform-thickness specimen, where advanced filtering or latitude techniques are not necessary, the procedure in setting-up a technique to obtain a good flaw sensitivity can be summarized as follows.

Procedure

1. Decide which technique (normal or critical) is required for the particular application under consideration.
2. Choose an appropriate film make and type.
3. From exposure curves make an approximate estimate of the kilovoltage required for a practical exposure time (usually within the range 0.25–4 min).
4. Use this estimate of kilovoltage required to determine, from Fig. 5.5 or Table 5.3, the film unsharpness U_f which will be present in the image. For example, for 270 kV X-rays the value would be 0.12 mm.
5. For 'normal' techniques the geometric unsharpness U_g is usually taken to be equal to twice the film unsharpness and so the appropriate minimum source-to-film distance is

$$\left(\frac{\text{focal spot diameter} \times \text{specimen thickness}}{2 \times (\text{film unsharpnes})}\right) + (\text{specimen thickness})$$

$$(6.10)$$

with all dimensions in the same units. Thus, for a 4 mm diameter focal spot, a 30 mm thick steel specimen and 270 kV X-rays, the source-to-film distance is

$$\left(\frac{4 \times 30}{2 \times 0.12}\right) + 30 = 530 \text{ mm}$$

This technique makes a compromise between a short exposure time and a small image unsharpness and is not suitable for applications where it may be required to detect narrow, fine cracks. For a better technique, the 'more critical' technique, the geometric unsharpness should be made equal to the film unsharpness, so the source-to-film distance is

$$\left(\frac{\text{(focal spot diameter)} \times \text{(specimen thickness)}}{\text{film unsharpness}}\right) + \text{(specimen thickness)}$$

(6.11)

and for the application described the distance is 1030 mm.

6. The intensifying screens used should be thin lead foil, 0.05 mm thick front screen and 0.1 mm thick back screen. Thinner screens can be used and the thickness is not critical. For X-rays below 120 kV, no front screen is necessary for intensification purposes, but a thin lead screen should usually be retained as a scatter-absorbing filter.

7. The film must be exposed to obtain a film density within the range 2.0–3.0 after normal processing. The film developer must be chosen to be suitable for the type and make of film used: most radiographic film developers are universal, but a few films and developers are incompatible.

8. Having now decided on film type, film density, source-to-film distance and kilovoltage, re-check from the exposure chart that the exposure time is still acceptable. If it is too long, choose a slightly higher kilovoltage and recalculate steps 1–7. In a few cases, such as with a thick specimen and equipment with a large focal spot, it may not be possible to use the 'more critical' technique without necessitating a long exposure time, and some additional compromises may be necessary. On very thin specimens and on light alloys, where low energy X-rays are used, the choice of kilovoltage becomes a more important factor. With gamma-rays the only factor controlling contrast which can be altered is the film, so the choice of film type and film density become most important. On very thick specimens, over 100 mm steel, contrast changes very slowly with X-ray energy, so again film contrast becomes extremely important.

Other technique factors

There are various other details of technique which must receive attention.

1. Identification letters and numbers (usually lead) must be placed on the specimen, so that they are shown on the radiograph.
2. An IQI should be placed on the specimen on a surface facing the radiation source. This must be of a type and size required by the customer: details of the best-known European types of IQI are given in EN-BS-462-1994.
3. Precautions should be taken against excessive radiation scatter. These include diaphragming the X-ray beam down to the minimum size to cover the film cassette; using a thick sheet of lead behind, or built into, the cassette; and masking the edges of the specimen. With small specimens there are filtering techniques which obviate such elaborate masking.
4. Ensure good film-intensifying screen contact.
5. Ensure adequate film viewing conditions appropriate to the film densities being used.

6.10 TECHNIQUES TO COVER A RANGE OF SPECIMEN THICKNESS (THICKNESS LATITUDE)

Not all specimens are of uniform thickness, nor is it always practical to take separate radiographs for each thickness change in a complex specimen. When a specimen having different thicknesses is radiographed with one exposure on one film, the thick parts will be shown as low densities on the film, and from what has been said of the parameters affecting flaw sensitivity, it is obvious that because film gradient varies directly with film density, there will be a loss of flaw sensitivity in these low densities, compared with a normal-density radiograph of these sections. At the other end the thin sections will be reproduced as high film densities, which soon reach the stage of being too high to be viewed.

There is therefore a range of thickness which can be examined on one film, limited at one end by the maximum usable film density, and at the other end by loss of flaw sensitivity.

There are several ways in which the thickness latitude can be extended.

1. By exposing the radiograph so that the thinnest section of the specimen has as high a film density as can be viewed with a bare Photoflood bulb, up to about 3.5.
2. Use a film with a lower film gradient, that is, sacrifice some contrast over the whole density range. For extreme cases a single-emulsion film can be used.
3. Shorten film development time.

4. Use a double film technique with two films of different speeds in the same cassette, exposed simultaneously.
5. Use higher energy radiation. This will lower flaw sensitivity in the thinner sections, but usually improves sensitivity in the thicker sections because of the increased film density.

It is not possible to specify the thickness latitudes which can be covered by each of these techniques, as the upper thickness limit depends on the poorest acceptable flaw sensitivity, but it is quite simple to calculate the range of thickness corresponding to any film density range, from an exposure chart and a film characteristic curve. Typical values for a film density range of 1.0–3.0 are

- 100 kV X-ray (aluminium alloys): 88–40 mm
- 200 kV X-rays (steel): 23–31 mm.

If a lot of work requiring latitude techniques is contemplated, it would be worth constructing a simple calculator showing the thickness range for a given density range for different kilovoltages, as shown in Fig. 6.11.

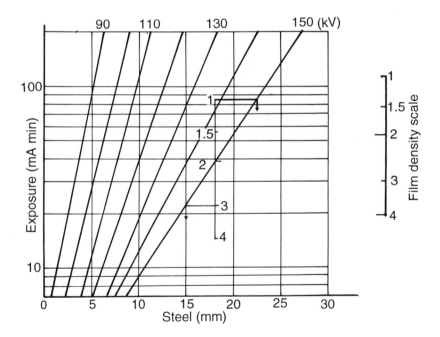

Fig. 6.11 Method of calculating thickness latitude. The scale of film density is placed across the exposure chart. In the example shown, an exposure of 39 mA min gives a density 2.0 on 18 mm steel, using 150 kV X-rays, and a density of 1.0 on 22.5 mm and 3.0 on 15 mm on the same radiograph.

6.11 SPECIMEN POSITIONING

There are a few general rules.

1. If two-dimensional defects are possible and their plane can be predicted, the beam of radiation should be directed along the probable plane of the defect. Thus, in some types of weld, lack of sidewall fusion is a possible and potentially serious defect, and the radiograph should be taken with the radiation beam along the face of the sidewall. Similarly, if there is a root-face, lack of root penetration is a possible defect and the radiograph should be taken with the beam along the root face. Lack of root penetration is rarely missed by radiography, but lack of sidewall fusion without associated slag inclusions is a more difficult defect to discover unless extra radiographs are taken as indicated.
2. The images of gas porosity and similar non-directional defects are virtually the same whatever the beam direction.
3. Whenever possible, a beam direction should be chosen which reduces the thickness latitude as much as possible.
4. The film should be located so that its mid-point is under the centre of the radiation beam.

6.12 MARKING: IDENTIFICATION

Every radiograph must show some permanent marks which identify the job, the region examined and the orientation of the film. Except on extremely thick sections of steel, thin lead letters and characters can be used and these, when placed on the specimen, whether on the film or source side, show clearly on the radiograph. They should be fixed on the specimen when it is a casting or weldment, and whenever possible left in position until the radiograph has been examined. Often fixing them with Plasticine is sufficient, but adhesive or tape is also used. At some later date the specimen will probably have to be permanently marked either by stamping, engraving or painting, according to what is most appropriate for the service conditions of the specimen. Typical lead characters for marking need only be about 4 mm high. It is undesirable to obscure the image of the specimen by using large quantities of thick and large lead letters, but in practice many more radiographs are spoilt by lack of identifying markers than by excess of characters. Whenever possible the markers should be put so as not to obscure significant image detail; thus on a welded seam, the characters may be put at the side of the weld seam and well clear of the heat-affected zone. Small plastic racks to hold a set of characters on a single line are available.

Similar markers are used to confirm overlap of films and to indicate specimen orientation, e.g. 0 and 90° on a cylinder.

Some characters such as A, O, T, U, V, W, X, Y, are symmetrical and can cause confusion when attempting to orient a radiograph; these should not be used singly if there is any chance of confusion. Perfectly adequate markers for many purposes can be made from small triangles of 3 mm lead sheet or snippets of lead wire.

In identifying small specimens, the lead characters should be placed on a shim of metal. If placed directly on the cassette surface they will be obscured by the very high background film density.

6.13 OTHER PRACTICAL CONSIDERATIONS

The unexposed film and the intensifying screens must, for use, be held in some device which protects the film from light, ensures good film–screen contact, and has a front surface which absorbs a minimum of the X-ray energy transmitted through the specimen; it must also be easy to load and unload in the darkroom. For most work a radiographic cassette is used, and these are commercially available in all the sizes for which film is supplied. Most commercial cassettes are based on medical designs, and tend to be rather fragile for industrial use. The cassette front is made of thin aluminium or of plastic and is easily damaged. Many cassettes incorporate a felt pad to provide even pressure on the film and screens and usually there is also a sheet of lead behind this to absorb extraneous scattered radiation which might otherwise be backscattered on to the film. Cassettes are available with curved surfaces (as well as flat) but in industrial work it is often necessary to have non-standard curvatures and sizes, and for these applications the film and intensifying screens can be contained in a plastic envelope, either a double envelope or a single envelope with a sealing flap. These have the limitation that good film–screen contact is more difficult to obtain. At one time evacuated envelopes and cassettes were available, from which, after loading the film and screens, the air could be pumped with a small hand-pump, so using atmospheric pressure to press the film and screens into contact, but these seem to have gone out of general use. Radiographic film is prone to pressure marks caused by local pressure, and plastic cassettes, if pressed against irregular surfaces, can lead to a crop of these spurious markings.

Some radiographic film is now marketed ready-packed in a strong paper envelope, together with lead intensifying screens. This type of packing is particularly convenient for seam-weld inspection where long lengths of narrow film can be used. The film required is about 80 mm wide and is purchased as a continuous roll, from which lengths are cut for use and the ends sealed. The lead screens are 0.02 mm thick and are bonded on to the inner surface of the 'paper' of the envelope. In some packaging the film and screens are packed under a vacuum, to ensure good contact.

Film processing

Radiographic films and developers are designed to obtain as high a contrast as possible, compatible with a reasonable film speed and fineness of grain. As already explained, processing consists of development, rinsing, washing and drying the film. Completely automatic processing units, in which the film is passed through the solutions by transport rollers, are widely used, as well as the conventional tank processing units in which a film is held vertically in either frames or clip hangers. With the latter type of processor a nitrogen gas bubbler is placed at the bottom of the developer tank to provide automatic developer agitation. Most automatic processors operate at a high temperature (27°–30°C), which requires special solutions. They also dry the films and the high-temperature processing permits short dry-to-dry times.

Spurious images

The interpretation of radiographs is discussed in Chapter 9. There are a number of spurious markings, commonly called 'artefacts', which can occur on films. The great majority of these are due to faulty processing and careless handling; a very small number are due to manufacturing faults. The more common artefacts are detailed in Chapter 9.

6.14 SPECIAL TECHNIQUES

A few special techniques will be mentioned at this stage. These are not used extensively, but are extremely useful on particular applications.

Multiple films

This technique has already been mentioned as a method of increasing thickness latitude. Several films are exposed in one cassette using an unmounted lead foil screen between each pair of films. If films of different speeds are used the faster film will give an image of the thicker areas of the specimen and the slower film of the thinner areas, one exposure giving the correct film density of about 2.0 on each film. Variants on this technique are to use two films of the same nominal speed, but alter the development time of one of the films.

   If two films are exposed together to give a low film density of (say) 1.0 on each film, they can be superimposed in registration for viewing and will have an effective density of 2.0. However, because of the shape of the film density characteristic, the effective contrast of the double film will be slightly less than the contrast of a single film exposed to have density 2.0; the exposure required for the double film will, however, be much smaller.

Stereoscopic methods

A pair of radiographs can be taken with the X-ray tube shifted laterally
in a plane parallel to the film between the two exposures. These radio-

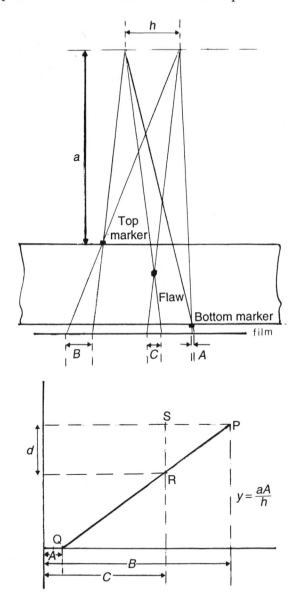

Fig. 6.12 Stereometric method of determining depth of a defect. A, B, and C are
measured on the film, and plotting these with $y = aA/h$ gives points P, Q and R.
Then RS = d is the depth of the defect.

graphs can then be examined in a stereoscopic viewer and with suitable markers placed on the specimen surfaces the position of a defect image inside the specimen can be seen, and its position in depth determined. Stereo images are not always successful with X- and gamma-rays, possibly because of the lack of reference detail, and a much more common technique is to make measurements on the films.

The principle of the method is shown in Fig. 6.12 and there are several variants according to whether the specimen thickness is accurately known or not. If the detail the depth of which is required to be measured is easily discernible the two required exposures can be made on one film, simplifying the method still further. If all the distances involved in the radiographic set-up can be measured, only the shift-distance of the defect image needs to be measured on the film, but it is more usual to put thin lead markers on the front and back surfaces of the specimen and also measure the image shift of these; the calculation is most conveniently made by a graphical method as shown in Fig. 6.12.

The simplest method of locating a defect, however, when the specimen is suitable, is to take two radiographs at right-angles with markers where necessary to show the edge of the specimen.

A number of very elegant stereoscope designs have been built for viewing stereo-pairs of radiographs, but very few industrial radiographs yield good stereo results. Part of the reason is probably psychological, and part because few of these stereoscopes will cope with films of the density used in industrial radiography.

REFERENCES

[1] Halmshaw, R. (1955) *J. Photo. Sci.*, **3**, 161.
[2] Klasens, H.R. (1946) *Philips Res. Rep.*, **1**, 2.
[3] Möller, H. and Weeber, H. (1964) *Arch. f. Eisenhutt.*, **35**(9), 891.
[4] Siiderow, A. (1979) *Eurotest Tech. Bull.*, **E 37**.
[5] Becker, E. (1964) *Materialprüfung*, **6**(2), 47.
[6] Berthold, R. (1938) *Atlas of Non-destructive Testing*.
[7] Newell, B.R. (1938) *Radiology*, **30**, 493.
[8] International Standards Organisation (1985) *ISO 5579*.
[9] CEN (1994) *EN.444*
[10] Deutsches Institut für Normung eV (1988) *DIN-54 111*.
[11] Li Yan (1994) *NDT&E Int.*, **27**(1), 15.
[12] Halmshaw, R. (1993) *Brit. J. NDT*, **35**(3), 113.
[13] Schnitger, D. and Mundry, E. (1975) *Materialprüfung*, **17**(7), 242.

APPENDIX: MEASUREMENT OF FOCAL SPOT SIZE

If the focal spot of the X-ray tube is greater than about 2×2 mm, its measurement is very simple. A series of holes are made with a fine needle in thin sheet lead of thickness 0.3–0.5 mm, so that the pinhole on the exit side of the needle is as small as possible, preferably less than 0.2 mm. The X-ray tube is then set up above a film at a distance of 1 m, and the sheet of lead with its pinholes positioned accurately midway between the focal spot and the film. The lead sheet and the film should be at right-angles to the centreline of the X-ray beam.

An exposure is made on fine-grain non-screen film to obtain a film density on the images of the spot(s) after processing of about 1.5; the background density need not be zero, if the spots are clearly seen. Only a short exposure time is needed, and a series of pictures should be taken to cover the normal working range of the X-ray set. The resulting images of the spots show the actual size and emission characteristics of the focal spot. It is unlikely to have sharp edges or be a uniform emitter.

In ASTM Standard E-1165-92, 'Focal spot measurement of industrial X-ray tubes by pinhole imaging', gives details with a recommended hole diameter of 0.03 mm for focal spots from 0.3–1.2 mm and 0.1 mm for larger spots. The image on the film is outlined and a correction factor of 0.7 is recommended to the axial dimension to produce an effective focal spot size.

The European Standards Organisation (CEN) has issued a standard giving six methods of measuring the effective focal spot size of an X-ray tube, ranging from a sophisticated 'pinhole' design in tungsten, to a microscanning system. Two of the CEN methods are for the measurement of microfocus tubes. One uses a small tungsten ball 1 mm in diameter, and the other a pair of crossed tungsten wires (1 mm diameter). These are imaged at very high magnification ($\times 20 - \times 100$) and the geometric unsharpness is measured on the image by microdensitometer traces. A similar method with a sharp metal edge is proposed for high energy X-rays. Some of these methods produce an unsharpness curve, and there are a number of proposals for assigning a single value to the unsharpness on the lines of Klasens' method (Fig. 6.6); the focal spot size can be calculated from this unsharpness.

7

Radiographic techniques: sensitivity

7.1 INTRODUCTION

A distinction must be made between radiographic quality and radiographic sensitivity. In many types of radiographic work the two are taken to be synonymous and this is particularly so in the field of flaw detection, where the ability to show a smaller flaw, i.e. an increase in flaw sensitivity, is regarded by the radiographer as an improvement in radiographic quality. However, in the inspection of irregular-thickness castings for example, ability to cover a range of thickness on a single radiograph with a reasonable sensitivity everywhere may be regarded as the criterion of quality and may be preferable to having maximum sensitivity in one particular thickness of the specimen. Thus, in this case, quality requires consideration of thickness latitude as well as sensitivity. Mechanisms and assemblies may be radiographed in order to measure some dimension such as a narrow gap, or the orientation of a component part, and then the most important criterion may be the definition of detail, sharpness of images of edges, or even correct angulation of the component relative to the radiation beam. In these cases radiographic flaw sensitivity may be a relatively minor factor in defining high-quality radiography.

However, a large proportion of radiography is concerned with flaw detection, and here sensitivity is an important consideration. There are limits to the dimensions of the flaws which can be revealed by radiography, and these limits and the factors controlling sensitivity will now be discussed.

7.2 DEFINITION OF TERMS

There are a number of terms commonly used in connection with radiographic sensitivity which are sometimes used in an ambiguous manner. It is desirable, therefore, to define the terms to be employed.

One of the oldest methods of measuring sensitivity is to place a series of thin plates of different thicknesses on the specimen; these plates are made of the same material as the specimen. A radiograph is then taken and the thinnest plate, the outline of which can be discerned on the film image, is taken as a criterion of sensitivity; the thickness of this plate is a measure of the sensitivity obtained on that particular specimen with the particular technique used. Radiographic sensitivity is often expressed in the form of a percentage of the actual specimen thickness:

$$\text{Sensitivity} = \frac{\text{thickness of thinnest discernible plate}}{\text{thickness of specimen}} \times 100\% \quad (7.1)$$

This number is called the 'percentage thickness sensitivity', or more usually the 'thickness sensitivity', and sometimes also 'contrast sensitivity'. Thus a plate 0.1 mm thick placed on a specimen 10 mm thick would be described as representing 1% thickness sensitivity. In the use of the term 'thickness sensitivity' it is implicit that the extra thickness is placed on the side of the specimen remote from the film.

Arising from this method of expressing sensitivity as a percentage it should be noted that if, with a different radiographic technique, a better sensitivity is attained, in the sense that a thinner plate can be detected, the numerical value of percentage sensitivity is smaller. For example, 2% sensitivity is not as good as 1% sensitivity if ability to show smaller detail is the criterion. Thus, such terms as 'higher sensitivity' and 'lower sensitivity' are ambiguous in that they may mean higher in the sense of higher quality, or higher numerically, which means the opposite. 'Better sensitivity' and 'poorer sensitivity' are less ambiguous expressions which should not be misunderstood.

The set of thin plates of different thicknesses, as described, could easily be made in the form of a small step-wedge in which each plate becomes one step of the wedge, and if the whole wedge is kept small it is frequently described as a penetrameter, or more specifically as a step-penetrameter or thickness penetrameter. The origins of this name are somewhat obscure. Medical radiographers originally used the term 'penetrometer' for an extra thickness of material which was laid across the specimen to confirm that some penetration of the specimen had been obtained (to check that all the blackening of the film was not due to scatter), and the name penetrameter apparently has penetrometer as its ancestor. Neither name is strictly correct as neither device measures penetration, and it is now accepted in most countries that the term 'Image Quality Indicator (IQI)' is a more acceptable term for devices which are put on the specimen to measure radiographic sensitivity. Such a simple form of indicator as the thickness penetrameter gives only a limited amount of information and there are many other designs of IQI consisting of wires, grooves, drilled holes etc., so that the simple step-type is almost obsolete. The method of specifying sensitivity, as judged with most of these devices, is the same as with the simple thickness penetrameter:

$$\text{Sensitivity} = \frac{\text{thickness of smallest visible element (wire, hole etc)}}{\text{thickness of specimen}} \times 100\%$$

(7.2)

and it is important to specify the type of IQI to which the value obtained applies, i.e. 'wire' IQI sensitivity or 'drilled-hole' IQI sensitivity. The IQI designs and the philosophy of these designs are discussed later; the immediate point is that 'penetrameter or IQI sensitivity have little or no meaning in a quantitative sense unless the design of the device and the method of reading the image are specified and understood. There are also IQI designs where the sensitivity obtained is quoted as a percentage which is not derived from equation 7.2, and this further emphasizes the ambiguities which can arise unless a precisely defined term is used. The use of the term 'percentage sensitivity' has proved to be so misleading that it should be avoided whenever possible. A much better method is to quote the actual thickness or diameter of the IQI detail which is perceived (or required to be perceived), i.e. wire or hole diameter.

The terms 'flaw sensitivity' and 'sensitivity of flaw detection' are also frequently met in radiography, and the term 'defect' is used synonymously with 'flaw'. The formulation of a general quantitative definition of either term is not possible, as flaws vary in shape and nature. Flaw sensitivity must not be confused with any form of IQI sensitivity, although it is obvious that some relationship between the two exists, in that an improvement in one is likely to result in an improvement in the other.

7.3 IMAGE QUALITY INDICATORS

An IQI may have a number of different purposes.

1. A common method of specifying radiographic techniques is to quote a value of IQI sensitivity that must be obtained for the radiograph to be regarded as of acceptable quality.
2. The sensitivity attained on different parts of a specimen, even on the same radiograph, may not be constant due to variations of thickness etc., and IQIs are used to assess the sensitivity in different parts of the image.
3. Even in closely-controlled techniques there can be unexpected and unpredictable variations which affect image quality. The importance

of any such effects needs to be judged, and an IQI is the method usually used.

4. There is a relationship between flaw sensitivity and IQI sensitivity, albeit a complex one (Chapter 8). Almost always, if a better IQI sensitivity is obtained, the flaw sensitivity will also be better.

The desirable characteristics of an IQI are therefore as follows.

1. It must be sensitive (in its readings) to changes in radiographic technique.
2. The method of reading the image should be as simple and as unambiguous as possible; different inspectors should obtain the same value from the radiograph; the sensitivity value obtained should be definite and precise. All these criteria call for a very careful choice of interval between elements, wires, holes etc.
3. The IQI must be versatile and easy to apply. Specimens for radiography are not of standard thicknesses and it is desirable that it should not be necessary to carry a large stock of different sizes of IQI. The IQI should be suitable for use on curved and rough surfaces.
4. The IQI should be small; the image has to appear on the radiograph and should not obscure useful parts of the image; the IQI image must not be capable of being mistaken for a specimen defect.
5. The IQI should preferably be cheap to make and to standardize.
6. The IQI should incorporate some means of identification of its size. Of these points, 1 is fundamental. From what has already been stated, it should be well-understood that sensitivity depends on a number of factors controlling contrast and definition. A good IQI must be sensitive to changes in any of these, but should not be unduly sensitive to contrast factors at the expense of definition factors, or vice-versa.

It is fair comment that no present IQI designs meet all these criteria, and most fall a long way short of ideal.

Types of IQI

In general, only the models of IQI designed for use with steel specimens will be described. In most cases there are exactly similar IQIs for use with other materials, with only a change in identification letters.

Wire-type IQIs

These consist of straight wires of the same basic material as the specimen. There have been a number of national standards, but during 1994 all the existing European standards for IQIs were replaced by EN-462-1(1994) which specifies wire lengths of either 10, 25 or 50 mm and wire

Table 7.1 Wire diameters, wire numbers for CEN wire
IQIs (BS-EN-462-1: 1994)

Wire number	Wire diameter (mm)	Tolerance (mm)
W 1	3.20	± 0.03
W 2	2.50	
W 3	2.00	
W 4	1.60	± 0.02
W 5	1.25	
W 6	1.00	
W 7	0.80	
W 8	0.63	
W 9	0.50	± 0.01
W 10	0.40	
W 11	0.32	
W 12	0.25	
W 13	0.20	
W 14	0.16	± 0.005
W 15	0.125	
W 16	0.100	
W 17	0.080	
W 18	0.063	
W 19	0.050	

diameters and wire numbers as shown in Table 7.1. The two thinnest
wires specified in the now out-of-date BS:3971 (0.032 and 0.040 mm) are
not included. The other wire diameters are the same. This is a standard
series of numbers of a geometric progression and the diameters correspond to British Standard wire gauges. The wires are placed parallel and
5 mm apart, between two sheets of low X-ray absorbent material such as
polyethylene sheet.

Earlier models using different wire diameters are now obsolete in
Europe. A series of models each containing seven wires from the series
is specified for iron, copper, titanium and aluminium. Each IQI should
have identifying symbols showing the wire material and the wire numbers used, as shown in Fig. 7.1(a). ASTM(USA) specified the same series
of diameters (0.0032 in–0.160 in) in E. 747–80, Part 11 (1982).

Step/hole types of IQI

In the same way, CEN Standard EN–462-2(1994) has replaced all European standards for the step/hole type of IQI. Table 7.2 gives the hole and
step dimensions and the step numbers. These IQIs consist of a series of
uniform-thickness plaques of the same material as the specimen, each
containing one or two drilled holes through the full thickness of the

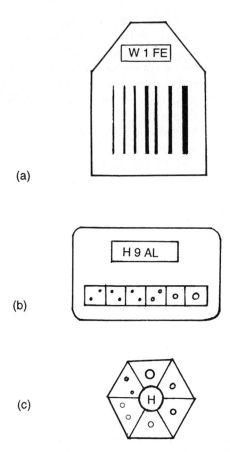

(a)

(b)

(c)

Fig. 7.1 Various types of IQI. (a) = wire type according to EN-462-1; (b) = step/hole type according to BS-EN-462-2; (c) = step/hole type according to BS:3971:1985.

Table 7.2 Hole and step dimensions for step/hole IQI

Step number	Hole diameter and step thickness (mm)	Step number	Hole diameter and step thickness (mm)
H.1	0.125	H.10	1.00
H.2	0.160	H.11	1.25
H.3	0.200	H.12	1.60
H.4	0.250	H.13	2.00
H.5	0.320	H.14	2.50
H.6	0.400	H.15	3.20
H.7	0.500	H.16	4.00
H.8	0.630	H.17	5.00
H.9	0.800	H.18	6.30

plaque at right-angles to the surface (Fig. 7.1(b)). The diameter of the hole is equal to the thickness of the plaque and is taken from the values shown in Table 7.2. The holes must not be bevelled.

In steps 1–8 there are two holes, not necessarily in the same orientation in each plaque, but not closer to one another or to the edge than 3 mm. In steps 9–18 there is a single hole in the centre of the plaque.

For convenience, the plaques may be machined as a series of steps on a single plate and a convenient step size is 12 mm. Typical designs are shown in Fig. 7.1(b) and (c). Each IQI has lead letters to identify it.

For use on ferrous materials, an IQI of low carbon steel is normally adequate and it is not necessary to have a special IQI of stainless, manganese, austenitic etc., steels. For aluminium alloys (of which there is a very wide range) it is desirable to match the material of the IQI to the alloy being examined when thin specimens are involved, but if there is insufficient application to justify purchase of a special IQI, one made of pure aluminium should be used. Comparison data have been produced between the sensitivities attainable using an IQI of pure aluminium placed on specimens of the same thickness of aluminium, and aluminium alloy containing 4% copper [1].

This step/hole design originated in France (AFNOR) and was also standardized, but not much used, in the UK. In the CEN standard, sets of six steps are chosen from the full series shown in Table 7.2.

American plaque-type penetrameters

In general, a different pattern of IQI is used in the USA and these are still called penetrameters. There are several minor variations of the same basic design. The best known is the ASTM design, which consists of a uniform-thickness plaque containing three drilled holes and identification letters (Fig. 7.2). If the plaque thickness is T, the hole diameters are T, $2T$ and $4T$ with an overriding limit that the minimum hole diameters are 0.01 in, 002 in and 0.04 in, respectively. The plaque must be of the same material as the specimen. Normally, a plaque is used such that T is 2% of the thickness of the specimen and the sensitivity is expressed in the form '(2–2T) level', which means that $T = 2\%$ and the $2T$ hole is discernible on the image.

'Equivalent penetrameter sensitivities' (EPS) are as shown in Table 7.3. According to ASTM Standard E–746–93, EPS(%) values are based on

$$EPS = \frac{100}{x}\left[\frac{T\,h}{2}\right]^{1/2}$$

where h is the diameter of the smallest detectable hole in a penetrameter step of thickness T with a specimen thickness x. E–747–80 gives conver-

Fig. 7.2 Example of ASTM plaque-type penetrameter

sion charts from wire IQI sensitivities to ASTM levels for different specimen thicknesses.

Compared with the European step/hole type of IQI, if a 2% thickness step with its corresponding 1T hole was found to be the limit of detection, the ASTM scheme would call this $(1-1T)$ or 1.4% 'equivalent sensitivity', whereas in Europe it would be 2.0% step/hole IQI sensitivity. With any step/hole IQI it is possible to have confusion between the discernibility of a step and the discernibility of a hole, but the two sensitivity values so obtained would be very different. Some US specifications distinguish between contrast sensitivity (judged on step discernibilty) and detail sensitivity (judged on hole discernibility).

Duplex wire-type IQI

A completely different pattern of IQI has been developed by the Central Electricity Research laboratorites (CERI UK) [2, 3] and was included in the British Standard on IQIs and their uses (BS: 3971: 1980). It is now

Table 7.3 ASTM levels of sensitivity (equivalents)

Level	Equivalent (%)
1–1T	0.7
1–2T	1.0
2–1T	1.4
2–2T	2.0
2–4T	2.8
4–2T	4.0

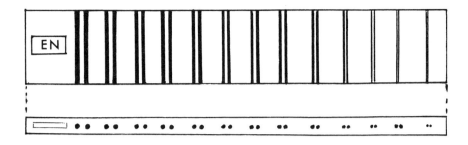

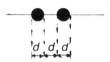

Enlarged view of each wire pair

Fig. 7.3 Duplex wire IQI, 70 mm long, 13 pairs of wires.

described in CEN draft Standard prEN–462–5 for the measurement of total image unsharpness values. It is particularly valuable for radioscopic work, and consists of pairs of straight wires made of very high density material (Pt, Au, W). Each pair of wires is spaced one wire diameter apart. The pattern is shown in Fig. 7.3 and the wire diameters and spacings are given in Table 7.4. The wires are mounted in a rigid plastic moulding and there are identification symbols in lead.

Table 7.4 Duplex wire IQI, element sizes

Element number	Wire diameter and spacing (mm)	Corresponding unsharpness (mm)
1	0.05	0.10
2	0.063	0.13
3	0.08	0.16
4	0.10	0.20
5	0.13	0.26
6	0.16	0.32
7	0.20	0.40
8	0.25	0.50
9	0.32	0.64
10	0.40	0.80
11	0.50	1.00
12	0.63	1.26
13	0.80	1.60

When placed on the source side of a specimen, the discernibility of the wire pairs is judged by eye, and the image of the first merged pair, that is, the first pair which cannot be seen as two separate wires, is taken as the criterion of the total unsharpness acting on the radiographic image, according to Table 7.4.

This type of IQI is difficult and expensive to fabricate, but nevertheless it provides information which is not available from the other more conventional patterns, and it also appears to be less prone to 'reading' errors. Originally, CERL (UK) also proposed a model B duplex wire IQI with bars instead of wires, with greater thicknesses, for use on specimens of 90 mm steel upwards, but this is no longer commercially available.

There have been many other proposed IQI designs. One type widely used in the UK at one time was the British Welding Research Institute (BWRA) step-wedge, in which a number symbol was formed in each step by means of small drilled holes. Grooved bars and rows of ball-bearings have also been proposed, as have lengths of micro-tubing. Recently, slits have been incorporated in a plaque-type IQI, but these have the disadvantage of introducing orientation factors, that is, the sensitivity obtained depends on the precise orientation of the IQI to the radiation beam.

Placement of the IQI

To be of any value the IQI must be placed on the source side of the specimen, and many specifications also ask for it to be placed near the edge of the diagnostic area of the image. If the IQI is placed on the film side of the specimen it would be nearer the film than any flaw in the specimen and would therefore be less influenced by factors affecting the blurring of the image. In this position the IQI is of very little value for judging image quality and may mislead the inspector. Placed close to the film, the IQI measures only image contrast, but in spite of this limitation some authorities believe that for applications where the correct location on the source side is inaccessible, an IQI near the film is better than no IQI image at all (section 7.5). Tables of attainable IQI values, including the case when the IQI is close to the film, are given in CEN.EN–1435:1995.

The Author's opinion is that if, due to specimen design, it is impossible to place the IQI on the source side of the specimen, it is probably best not to use an IQI but to check the technique to be used on a mock-up specimen of the same thickness and geometry. The IQI should be placed on a thickness of material which matches the important part of the specimen in thickness. It is often recommended that a wire IQI be laid across the weld or that, when a weld with reinforcement is examined, a shim of metal should be placed under the IQI to match the total thickness through the weld. On a specimen of varying thickness it is desirable to use more than one IQI (Section 7.5).

7.4 FILM VIEWING CONDITIONS

To assess IQI sensitivity the radiograph is placed on an illuminated screen of appropriate brightness (luminance) for the film density, with the film correctly masked to eliminate glare emanating from around the film or from any part of the film with a particularly low density. The diameter of the smallest wire or drilled hole which can be detected with certainty is taken as a measure of the IQI sensitivity attained. Good film-viewing conditions are absolutely critical, as it is quite possible to have information on a radiograph which is not seen because of too high a film density or too low an illuminator luminance.

The importance of relating illuminator luminance to film density was recognized by the International Institute of Welding, whose Commission VA produced detailed recommendations (IIS/IIW-335-69), which later became ISO:5580 and then BS-EN-25580:1992. These standards state that the luminance of the illuminated radiograph should not be less than $30\,cd\,m^{-2}$, and whenever possible approximately $100\,cd\,m^{-2}$ or greater. This minimum value requires illuminator luminances for

- film density 1.0: $300\,cd\,m^{-2}$
- film density 2.0: $3000\,cd\,m^{-2}$
- film density 3.0: $30000\,cd\,m^{-2}$.

These values have been based on experiments which directly examined the discernibility of IQI images on radiographs of different densities at different illuminator luminances: the results obtained are confirmed by the classical work of König [4] on the discernibility of small contrasts at differing luminances.

The practical difficulties of obtaining such high luminances for films of density greater than 2.5 have caused some authorities to use a lower acceptable minimum luminance of $10\,cd\,m^{-2}$ instead of $30\,cd\,m^{-2}$: it is argued that this will permit films of density greater than 3.0 to be accepted and the gain in contrast thereby will counterbalance the loss in observed contrast due to the lower image luminance. The colour of the viewing light should preferably be white and the light must be diffused, but need not be fully diffuse. The radiograph should be examined in a darkened room or enclosure, with care being taken that as little light as possible is reflected off the film surface directly towards the film reader. The luminance of a white card put in place of the radiograph should not exceed 10% of the luminance of the illuminated film.

The IIW recommendations also suggest that for anyone coming from full daylight, some dark adaptation is necessary before commencing viewing, and also that some check of the film reader's eyesight at close-viewing distance is necessary. The use of a low power ($\times 2 – \times 4$) magnifying glass is recommended for the study of fine detail on the radiograph.

All this may seem pedantic, but practical experiment with a group of film inspectors has shown that person-to-person variations in readings made on the same radiograph are a very significant factor in trying to determine attainable values of IQI sensitivity. A second cause of human variation in IQI readings is the phrase used earlier, 'can be detected with certainty'. If a radiograph of a row of wires of progressively decreasing diameter is examined, there is no sudden point at which the image of a wire disappears. The images gradually become less easily discernible, then discontinuous, and finally undiscernible. However, it is necessary to set oneself a personal standard of what one means by 'detected with certainty'; to judge from cooperative experiments, different film readers can set themselves very different standards. With the image of a wire there is a stage when the image begins to appear discontinuous, but one is quite certain that a wire exists. Some workers have expressed the standard of discernibility as a percentage probability. The opinion of the Author is that if part of the wire can be discerned it should be counted, but discussions with other radiologists make it clear that this criterion is not universally adopted.

7.5 CODES OF RECOMMENDED GOOD PRACTICE

Many countries have issued codes and standards specifying good practice in film radiography, detailing and usually explaining the technique parameters required. The pioneers were the British Standards Institution, the American organisations. ASTM and ASME, and the German DIN. Similar codes have been published by the IIW and ISO.

In 1992 the EEC decided to redraft all the European standards on NDT and have a CEN set of standards. Many of these have now been issued and they replace the national standards in all the EC countries. They have been accepted by majority voting although some countries have opposed some of the details.

The existing CEN, ISO and US standards on radiography are listed in Table 7.5. Some of these are currently being revised and some of the CEN standards are still in provisional form. In addition, CEN is preparing drafts on focal spot size measurement, radioisotope source size measurement, X-ray kilovoltage measurement and radioscopy.

Non-EC countries which based their standards on BS or DIN documents are of course not mandated to change, and some large purchasing organizations such as the oil companies, the electricity and gas boards and UKAEA have their own standards.

Table 7.5 Major standards concerned with radiographic techniques

Standard	Description
CEN	
prEN 1330	Terminology (vocabulary of terms)
EN-444:1994	General principles for the radiography of metallic materials with X- and gamma-rays
EN 584–1	Film classification
prEN 584–2	Control of film processing by reference strips
EN 462 Parts 1–6	Image quality indicators, and their uses
prEN-1435: 1995	Radiography of welds
ISO	
ISO: 5579	Radiographic examination of metallic materials with X-and gamma-rays: basic rules (under revision, 1995)
ISO: 5580	Film viewing conditions
ISO: 2437	Recommended practice for the X-ray inspection of fusion-welded butt joints for aluminium and its alloys and magnesium and its alloys, 5–50 mm thick (1972)
ISO: 2504	Radiography of welds and viewing conditions for film; utilization of recommended patterns of IQI (1973)
ISO: 3777	Radiographic inspection of resistance spot welds for aluminium and its alloys – recommended practice (1976)
USA	
ANSI Z166.7, 1972	Method for controlling quality of radiographic inspection
ANSI Z166.22, 1974	Recommended practice for radiographic testing
MIL STD 00453A, 1975	Radiographic inspection

pr: for CEN standards this indicates a provisional standard

The philosophy of radiographic standards varies considerably from one country and organization to another. Some detail all the technique parameters, others specify an IQI value which should be obtained and leave the details of the technique to the radiographer. American philosophy is generally different from European in that in some standards an acceptable sensitivity is quoted in terms of the use of the ASTM plaque-type penetrameter and no other detail of technique is specified. Some standards detail a range of techniques such as 'very high sensitivity', 'high sensitivity' or 'medium sensitivity' and leave the user to decide which of these to employ for any particular application.

An alternative approach is to ask a radiographic department to submit a specific technique and sample radiographs to the customer or the customer's inspector: this technique, if acceptable, is then 'sealed'. Some standards and codes of good practice specify detailed IQI sensitivities in addition to technique details, and in the newer codes these sensitivity values are reasonably realistic. Some standards also specify acceptable and rejectable levels of defects.

Butt welds in plates

British practice has been to write separate standards for different specific applications, for example:

- butt welds in steel plate from 5–50 mm thick;
- butt welds in steel plates from 50–500 mm thick;
- butt-welded circumferential weld joints in steel pipes;
- fusion-welded branch and nozzle joints in steel pipes.

whereas the main German Standard (DIN-54-111) and the CEN Standard, now BS-EN-444: 1994, 'General principles for radiographic examination of metallic materials by X- and gamma-rays', attempt to cover all metallic materials and thicknesses. Most of the clauses in EN-444: 1994 are repeated in the CEN Standard prEN-1435 for weld inspection which has additional clauses on pipe-weld set-ups and required IQI sensitivity values. EN-1435 was written to cover all metallic welds. Both these CEN standards detail only two classes of technique:

- Class A, basic techniques;
- Class B, improved techniques ('Will be used when Class A may be insufficiently sensitive').

But it is also stated that 'better techniques compared with class B are possible and may be agreed between the contracting parties'. In general, class B uses a larger source-to-film distance, a higher class of film and a higher minimum density than class A. Higher class radiography may take a little longer but the overall inspection cost is not markedly increased.

Circumferential pipe welds

As circumferential pipe butt welds on which radiography is necessary can range in size from 10 mm internal diameter to several metres internal diameter, this is a much more complicated application of radiography and there is more controversy over appropriate techniques. The larger pipes can obviously be treated as plate welds and it may be possible to use radiographic equipment inside the pipe. On many pipe weld inspections, however, there is no easy access to the inside, except by using remote-controlled crawlers to carry the radiographic equipment along the pipe from an open end, which may be hundreds of metres away from the weld. The use of crawlers as such has not been the subject of any national or international code of practice.

There are three basic arrangements for the radiography of pipe welds, which are detailed in all standards, including prEN-1435. These are described below.

Source of radiation outside, film inside

This method has the advantage that the film is on the side of the weld where the most critical defects (root cracks etc.) occur (Fig. 7.4). There is, therefore, an increase in the detection sensitivity for these defects due to the smaller geometric unsharpness. On the other hand, the film is curved away from the source of radiation, so that the length of weld which can be examined at each exposure is rather small and the geometry at the edge of each film results in some image distortion. In practice, at least six

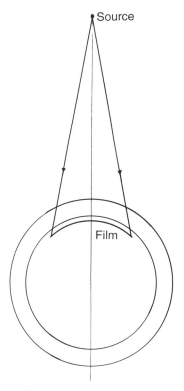

Fig. 7.4 Pipe weld radiography: source outside, film inside.

or more radiographs are necessary to cover a complete weld and the source of radiation has to be moved for each exposure; on some specifications with stringent density rules, 12–14 films may be needed per weld. prEN-1435 gives nomograms showing the number of films required in relation to wall thickness and pipe diameter. This method is slow, and more suitable for large cylindrical vessels which can be rotated than for pipe welds.

Source of radiation inside, film outside

This is the most economic method, if it is possible to get a suitable source of radiation inside the pipe. If the source is placed at the weld centre (Fig. 7.5 (a)), the length of weld which can be examined at one exposure is limited only by the size of the radiation field: with a gamma-ray source, or an X-ray tube with a conical target, the whole weld can be covered in one exposure. The method is so attractive that special equipment (crawlers) are made to carry an X-ray unit or a gamma-ray source inside a pipe to the correct location. A gamma-ray source is very convenient to use in this manner, if the pipe-wall thickness is appropriate. With small diameter pipes, which allow only a small source-to-film distance, a very small diameter source of radiation is essential to obtain good image definition. Thus, using 150 kV X-rays on a 60 cm diameter pipe, the effective diameter of the focal spot would need to be 0.6 mm, assuming a wall thickness of 12 mm. If this requirement is too restrictive, the radiation source can sometimes be placed off-centre (Fig. 7.5(b)). In this

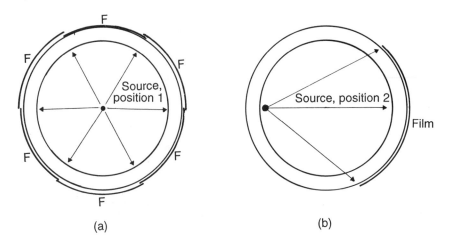

(a) (b)

Fig. 7.5 Pipe weld radiography: source inside, film outside. (a) = source at centre position, film wrapped outside (film may be one continuous length or separate lengths); (b) = source offset.

case about four exposures are needed to cover the whole weld, but this figure needs to be worked out for each (pipe diameter/wall thickness) ratio.

On small diameter thin-wall pipes the use of a gamma-ray source of Yb-169 on the centreline of the pipe is an important new development, as Yb-169 sources become more generally available. Such a source can often be positioned with a miniature crawler and special small-diameter spherical sources of Yb-169 are available, so that the geometric unsharpness is the same whatever the orientation of the source.

Source of radiation outside, film outside

This method must be used when there is no access to the inside of the pipe. It has a number of variants according to the diameter of the pipe. If this is of large diameter, the radiation source can be placed close to one wall with the film on the opposite side, as shown in Fig. 7.6(a). If this would result in too short a source-to-film distance the source must be stood off the pipe, as shown in Fig. 7.6(b). In both cases the image of the part of the weld nearest the source will be too blurred to be of any use and the technique is called 'double wall: single image'. If, however, the pipe diameter is small and the source-to-film distance large (Fig. 7.6(c)),

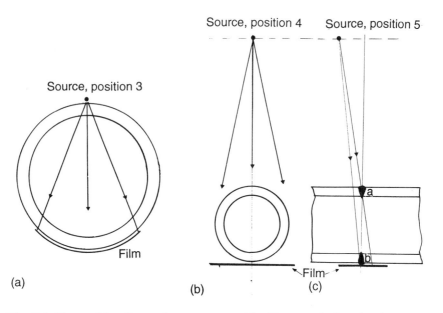

Fig. 7.6 Pipe weld radiography, source outside, film outside (inside of pipe not accessible). (a) = double wall, single image. (b) = double wall, double image; (c) = double wall, double image, showing source offset along axis of pipe.

the images of the parts of the weld at both a and b may be sharp enough to be useful and the technique is then known as 'double wall: double image'. In this case the source is often offset from the plane of the weld, otherwise the images of the two parts of the weld will be superimposed and may be confused. This offset of the source produces an elliptical image of a circular weld; the offset distance should be kept to a minimum, and one-fifth of the source-to-film distance is often recommended. It has been found in practice that if an excessive offset distance is used, narrow root defects can be missed or become more difficult to interpret, and there is a recent tendency to use little or no offset.

In BS 2910: 1973, the 'double wall: double image' technique is recommended to be used only on pipes less than 90 mm in diameter. Whenever possible the film should be bent to fit close to the weld surface to reduce the defect-to-film distance to a minimum; for very small pipes a flat cassette is often used, but there appear to be no definite recommendations about this in current standards as several factors are involved. Thus, if one were radiographing a 75 mm pipe with an focus-to-film distance of 120 cm, a flat cassette would be acceptable; if the focus-to-film distance was 50 cm, such a cassette would not be satisfactory.

With the double wall: double image technique generally only two exposures, taken mutually at right-angles, are made. With the single image method a minimum of four exposures is usually recommended, although some codes are not specific on this point. The limits to the length of weld to be examined at each exposure is set by the fall-off in film density between the centre and edge of the film. Typically, a density range of 1.5 to 2.5 or 3.0 is permitted, but some standards specify a smaller range, and on some pipes the limit may also be set by the distortion of peripheral images. If the source of radiation can be placed close to one side of the pipe, it can be placed virtually on the weld, with no offset, without interfering images being produced.

In all double-wall techniques the radiation must penetrate two walls and this reduces the flaw sensitivity compared with a single-wall technique. For this reason, if for no other, double-wall methods should not be used unnecessarily. Much controversy can arise when one tries to compare the relative merits of using a gamma-ray source inside (single wall) with a double-wall technique with X-rays outside, and none of the existing standards give any general rules on this problem. It has been shown that on an 18 mm thick steel pipe, a gamma-ray technique using Ir-192 inside the pipe can give better IQI sensitivities than X-rays with the double-wall method.

No existing code of practice offers much guidance on the relative performances, in terms of flaw sensitivity, of the three basic techniques shown in Figs 7.4–7.6. The crux of the problem is that, assuming that placement of a source at position 1 in Fig. 7.5 will cause an unacceptably large value of U_g, a comparison is needed between having the source in positions 2, 3 or 4 in Figs 7.5 and 7.6. At position 2, when this is physically possible, the radiation has to penetrate only one wall. Consequently, there is good image contrast, but unless a very small diameter source is available, there may still be too large a geometric unsharpness. At position 3 there is the advantage of a short exposure time and no interference from the image of the upper part of the weld, but again there may be unsharpness problems unless a small diameter source is available, and the exposure times will not be as short as at position 2. The radiation has to penetrate two walls so there will also be a loss in contrast compared with position 1. At position 4, image definition can be as good as required but the source must be offset to separate the two images, and for a fixed weld the practical problems of source positioning around the weld are difficult. Again, because the radiation has to penetrate two walls there will be a loss of image contrast. The choice should be position 1 on Fig. 7.5 whenever this is possible and when a suitable small source is available: the recent development of small-diameter Yb-169 sources has eased this difficulty considerably on thin, small-bore pipes. It should be noted that if the source is placed at position 1, the geometric unsharpness will be approximately doubled compared with position 2, but the exposure required will be about one-quarter. If, however, the source diameter has to be halved to retain the same geometric unsharpness, this would mean that the source volume, and presumably the source intensity, will be one-eighth, so the exposure would need to be doubled.

On double wall: single image inspection, the specification of IQI sensitivity as a percentage value has caused much confusion. In prEN-1435 this problem has been avoided by having tables. Six tables of required IQI sensitivities in terms of the minimum wire or hole diameter which must be seen on the radiograph are provided for both wire IQIs and step/hole IQIs. Due to the tabular, discontinuous nature of the data, these are not entirely logical, but broadly the change from a single-wall technique to a double-wall technique causes a loss of between 1 and 1.5 wires in wire sensitivity. Placing the IQI on the film side instead of on the source side causes an apparent increase in wire IQI sensitivity of one wire (Chapter 8).

The precise positioning of an IQI causes considerable difficulty when writing standards and specifying acceptable IQI values. Assuming the weld has not been dressed (i.e. has not been ground smooth to the parent plate surface), the thickness through the weld (t_1 on Fig. 7.7(a)) may in practice be 50% larger than the pipe wall thickness t_0, so an IQI placed at R will be on a lower density region of the radiograph compared with position Q and so give a considerably poorer sensitivity value. If the value is expressed as a percentage of the penetrated thickness at R, the value will be even poorer.

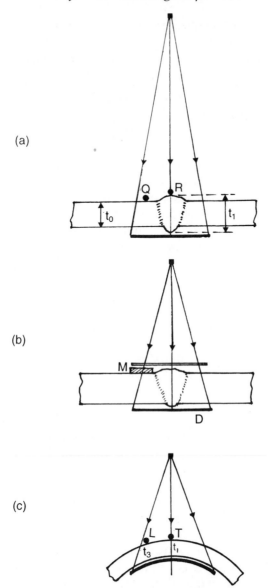

Fig. 7.7 Pipe weld radiography. Positioning of the IQI.

It is often recommended that the wires of a wire IQI are laid across the weld (Fig. 7.7(b)), and if the weld has had the reinforcement left on, on either surface, the same problem arises: the image of the wire under the thicker metal disappears while still seen at D. The use of a compensating-thickness shim M is sometimes recommended and eases the problem, but in practice this is rarely done.

Table 7.6 Thickness range in steel on which gamma-rays may be used for weld radiography

Source	Penetrated thickness (mm)	
	Test class A	Test class B
Yb–169	1–15	2–12
Ir–192	20–100	20–90
Co–60	40–200	60–150

In the case of 'film inside, source outside' (Fig. 7.7(c)) in pipe-weld radiography, the penetrated metal thickness at t_3 is greater than at t_1, so the film density is lower, even if the weld has been ground smooth, so putting the IQI at the edge of the diagnostic area of the film (L), as is often recommended, produces a poorer IQI sensitivity value than if the IQI is at T. In addition, on very small pipes, it is not really practical to put the IQI anywhere but centrally.

Another major difference between CEN-EN-444-1994 and older national standards is the minimum pipe-wall thicknesses on which gamma-rays are permitted. Table 7.6 gives the CEN proposals (at the time of writing these values were not consistent through various CEN standards).

The case of thin-wall small-diameter pipe weld inspection is not met very adequately in any current national or international code of practice. The best techniques use either very small Yb-169 sources with very short source-to-film distances, or a microfocus rod-anode tube with the focal spot on the centreline of the weld. Owing to the very short source-to-film distance, when the gamma-ray source is on the centreline of the weld, even with a small-diameter low-intensity source exposure times are likely to be short (Fig. 7.8). With a Yb-169 source, some workers have recommended a no-screen technique for specimen thicknesses of steel less than 5 mm. For steel thicknesses greater than 5 mm, lead intensifying screens should be used, with a 0.02 mm front screen. On specimens less than 5 mm steel, a significant loss of detail sensitivity was noted [5,6] when using thin lead intensifying screens, which is presumably associated with the filtering effect of the specimen and front screen on the initial radiation spectrum. With thicker specimens this effect almost disappears. As thin lead screens are claimed to produce an intensification factor of six, their use has the advantage of shorter exposure times, which is particularly important if double-wall techniques are used.

On pipe welds less than 25 mm outer diameter a flat cassette can be used and the tangential image of parts of the weld at the two sides where the weld-root profile is seen can be quite useful.

With a ytterbium-169 source a thickness range of 25–55 mm for test class B in aluminium and titanium has been proposed (EN-444-1994).

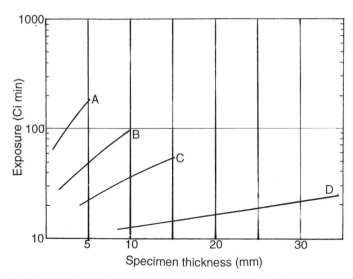

Fig. 7.8 Radiography with a ytterbium-169 source: MX film; automatic processing; film density 2.5; 120 mm source-to-film distance. Curve A = no intensifying screens, steel specimen; Curve B = back screen only (lead), steel specimen; Curve C = thin lead intensifying screens, front and back, steel specimen; Curve D = thin lead intensifying screens, front and back, aluminium alloy specimen.

The use of very short source-to-film distances requires that great care is taken to maintain the specimen-to-film distance at a minimum, to keep the geometric unsharpness as small as possible. Thus, for a 2 mm steel weld, the minimum source-to- film distance is recommended to be 24 mm for a 1 mm diameter source, based on $U_g = 0.09$ mm, and if the film is allowed to stand-off 2 mm, the real value of U_g would double to 0.18 mm.

The upper limits of thickness in Table 7.6 for Yb-169 gamma-rays seem high, and for thicknesses above 10 mm steel an Ir-192 source may give better results. A microfocus rod-anode X-ray set will give still better results by producing higher contrast radiographs, but if there are material thickness irregularities such as weld reinforcement, the radiographs may be more difficult to interpret.

Castings

Codes of radiographic practice are also available for the radiography of castings, but these are more in the form of guidance notes than precise instructions because castings are diverse in size, shape and materials. CEN-EN-444 (General principles) covers many of the points, with an emphasis on class A for general castings and class B techniques for aerospace components. Other CEN standards are in preparation.

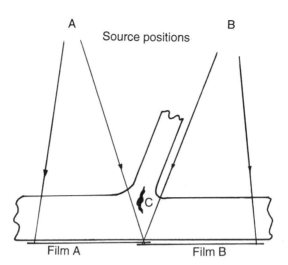

Fig. 7.9 Casting radiography: overlap of areas covered by two films. As shown, a serious shrinkage cavity at C would be missed by the two radiographs.

In general, flaws which are markedly directional are less common in castings, and the main problem in the radiography of complex castings is to ensure complete coverage; flaws tend to occur at section changes where coverage by radiography is most difficult.

A second common difficulty is to ensure that radiographs of two adjacent areas do completely cover the areas. Figure 7.9 illustrates a simple example, where the two radiographs A and B could fail to reveal the shrinkage cavity at C. On any important casting the designer and founder should both be consulted, the first to determine the most critical areas from the stressing viewpoint, and the second on the likely regions for casting flaws to occur. A knowledge of the casting procedure or some advice from the foundry is of great value in deciding on beam orientation, particularly on large castings where the radiography must be selective to be economic.

On small high-duty castings, where it is required to cover all the casting with radiographs of good sensitivity, there are no general rules. The usual recommendations for good radiographic practice should be applied:

1. penetrate the minimum possible thickness;
2. try and keep the film close to the metal surface;
3. remember the need for masking edges etc. when using X-rays;
4. use as high a film density as possible on the thin regions, when covering a thickness range;
5. consider 'thickness latitude' techniques, e.g. filters, double films;

6. take special care to cover corners and section changes, with a suitably angled radiation beam.

It is easy to put lead markers on both surfaces of the casting and to include a number of IQI, and from the images of these to determine exactly what coverage has been obtained and what sensitivity.

The only common, very narrow directional flaw in castings is the stress crack which almost always reaches the surface and so should be discovered by a surface crack-detection examination; such a surface examination should always complement casting radiography. Hot tears, the other type of 'crack' which occurs in castings, twist and turn so violently that they are rarely missed by radiography.

7.6 SPECIAL TECHNIQUES

Megavoltage X-radiography

Many of the points of technique relevant to radiography with high energy X-rays have already been covered, but a few points require special emphasis.

Some aspects of equipment performance such as field size, which are of only secondary importance with low energy X-rays, become much more important with high energy X-ray equipment and other factors such as focus size become less important. In addition to radiography on film, there is also increasing interest in television–fluoroscopic techniques, using high energy radiation. Qualitatively, the effect of increasing the X-ray energy above about 1 MeV is:

1. to increase the probability that the scattered radiation will be in the forward direction, i.e. at a smaller angle to the direct beam;
2. to reduce the proportion of scattered-to-direct radiation reaching the film through a given thickness of material;
3. to increase the amount of radiation which penetrates a given thickness of material.

For X-rays of energy less than 10 MeV, Compton absorption is the predominant absorption process, so far as steel is concerned; at higher energies, pair production becomes relatively more important, but for the energies used to date in radiography this is not likely to be as important as Compton absorption. It is the combination of 1–3 above which form the attractiveness of radiography with high energy X-rays and offers the possibilities of penetrating large thicknesses of material with useful values of defect sensitivity.

The equation for thickness sensitivity (equation 2.18) contains the ratio $[1 + (I_S/I_D)]/\mu$, which represents the effect of the radiation energy (sometimes called the 'radiation contrast factor'); a decrease in the value of this ratio signifies an improvement in thickness sensitivity. In Chapter 8 it is also shown that detail sensitivity, both with IQI and defects such as cracks, depends on the same factor. A study of the variation of this ratio for different X-ray energies and different thicknesses of material should therefore give information on the choice of the most suitable X-ray energy for different thicknesses.

It is not possible to calculate I_S/I_D to the required degree of accuracy for radiographic conditions for heterogeneous X-ray beams, and it is necessary to use experimental data. The results obtained lead to a remarkable conclusion. They show that the thickness sensitivity, that is, the contrast, changes with X-ray energy in a totally different way with specimens of 70 mm steel and thicker, compared with thinner specimens. For a 25 mm steel specimen an increase in X-ray energy results in a higher value of $[1 + I_S/I_D]/\mu$ and so a poorer value of thickness sensitivity. This is a well-known effect: too high an X-ray energy gives a poor radiograph. However, for very thick specimens, where one must use megavoltage X-rays to obtain penetration the effect is reversed; for a 100 mm steel specimen, a better thickness sensitivity can be obtained with 10 MV X-rays than with 1 MV X-rays. The reason is that $[1 + I_S/I_D]$ is decreasing more rapidly than μ, so that the ratio gets smaller as the X-ray energy is increased. If, therefore, this is the only effect of X-ray energy on radiographic sensitivity, there is considerable reason to use very high energy X-rays for thick specimen radiography; sensitivities can be better and exposure times will be shorter.

The problem is not, however, quite so simple. One is concerned with detail sensitivity, and the influence of X-ray energy on the definition attainable must also be considered. Radiographs taken with high energy radiation do not have the same sharpness of detail as those taken with lower energies, so that increase in film unsharpness must also be taken into account (see Table 5.3 for U_f values).

The most popular megavoltage X-ray equipments are linacs in the 4–8 MeV range. This is governed partly by the increasing cost of building still higher-energy machines, partly because the output of an 8 MeV linac is more than adequate for most applications, and partly because as one increases the energy the field size becomes more restricted.

Considering 8 MV X-rays from a linac, the output is likely to be of the order of 2000–6000 R min^{-1} at 1 m distance. The field size will be 1 m diameter at 2 m distance and the focal spot size 2 mm diameter. The film unsharpness with this energy of radiation is 0.6 mm and an exposure chart for steel is shown in Fig. 7.10. For a specimen of 250 mm steel thickness, radiographed at 3 m source-to-film distance on MX film, the geometric unsharpness will be 0.17 mm and the exposure time less than 0.5 min. Note that with linacs the X-ray output is usually measured in

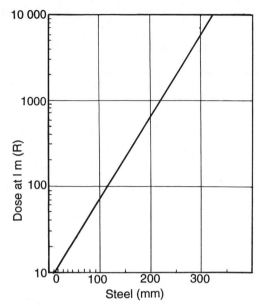

Fig. 7.10 Exposure data for 8 MV X-rays: 1 m source-to-film distance; MX film, density 2.0; copper intensifying screens, 1.5 mm front, 0.25 mm back.

radiation units and not in milliamperes, and the source-to-film distance is chosen in terms of the required field size rather than U_g.

With some linac designs the electron beam can be refocused after acceleration, so that the electrons are brought on to the target at a considerable angle to the normal. This modifies the distribution of X-ray intensity across the X-ray field. As an example, with one 5 MeV linac the field size (half-angle to half-intensity) was increased from 11° to 27°. The very large radiation output of modern linacs not only enables large values of source-to-film distance to be used, but also makes it possible to use fine-grain or even ultra-fine-grain films for almost all applications.

For intensifying screens, however, special materials are necessary to obtain the best results. Conventional lead intensifying screens can be used, but better sensitivities can be obtained with different screens [7]. Table 7.7 shows the best screen materials and combinations for use with different energies of radiation.

Most of the image degradation comes from the back intensifying screen. With the thick screens recommended for use with 3 MV X-rays and higher energies, great care must be taken to ensure that good overall film screen contact is obtained.

With 8–30 MV X-rays the best results are obtained with no back inten- sifying screen, but if there is likely to be much backscattered radiation

Table 7.7 Intensifying screens for use with megavoltage X-radiation

Equipment radiation	Useful thickness range, steel (mm)	Screen material	Front screen thickness (mm)	Back screen thickness (mm)
1–3 MV X-rays	50–200	Lead	1.0–1.6	1.0–1.6
3–8 MV X-rays	70–300	Copper	1.0–1.6	1.0–1.6
8–30 MV X-rays	70–400	{ Tantalum Tungsten	1.0–1.5	None
Cobalt-60 gamma-rays	50–150	{ Copper Steel	0.5–1.0 0.2–1.0	0.5–1.0 0.2–1.0

generated behind the film cassette, a back screen may have to be used to absorb this. The main purpose of the front screen, with high energy radiation, is to absorb some of the softer radiation generated in the specimen rather than produce any intensification.

With cobalt-60 gamma-rays, copper or steel screens will produce considerable improvement in detail sensitivity compared with lead screens, but the exposure needs to be doubled.

These relationships between exposure time, sensitivity and field size for steel specimens for megavoltage radiation are summarized below.

1. For thicknesses around 100 mm steel, radiographs can be obtained with 2 MV X-rays with reasonable exposure times and virtually no practical restriction on film size. The sensitivity attainable can be as good or slightly better than will be obtained with any higher X-ray energy.
2. For thicknesses around 200 mm steel, 5 MV X-rays or higher energies are needed if exposure times are to be kept short. If large field sizes are required the more powerful linacs have many advantages, but if small fields are adequate, betatrons require only reasonably short exposure times.
3. For thicknesses around 300 mm steel, only the large linacs will permit short exposure times on fine-grain film.
4. On 100–300 mm steel, the differences in attainable sensitivity between 5, 8, 20 and 30 MV X-rays are comparatively small.
5. On 25–70 mm steel, 1 MV X-rays will give sensitivities considerably better than higher energy X-rays.

As betatrons usually have a focal spot which is less than 1 mm wide, projective magnification can be successfully used and this is frequently given as one of the major advantages of a betatron, compared with linacs. Projective magnifications of 2:1 and even 3:1 have been used and improvements in sensitivity, particularly in the detection of very narrow cracks in line with the X-ray beam, have been reported.

Projective magnification methods

The advent of microfocus X-ray tubes has led to an expansion of interest in projective techniques. Reliable and robust microfocus X-ray tubes are now available operating up to 200 kV or even 300 kV, with effective focal spot sizes down to 10 μm or smaller.

By spacing the film away from the specimen the image of detail in the specimen is projected larger than natural size on to the film and so the effect of film grain is less marked; because the X-ray tube focal spot is very small the geometric unsharpness is not increased sufficiently to produce significant image blurring. There is obviously an optimum magnification above which no further improvement will be obtained. Also, the technique should be compared with radiography on a single-emulsion film with the specimen close to the film, when such a radio-graph is examined by optical magnification.

The second advantage of projective methods is that for magnifications greater than about × 3, the film is so far from the specimen that there is virtually no scattered radiation generated in the specimen reaching the film, that is, $(1 + I_S/I_D) = 1$.

From the established geometric relationships:

$$M = \frac{a + b}{a} = 1 + \frac{b}{a} \qquad\qquad (7.3\,a)$$

so

$$U_g = \frac{sb}{a} = s(M - 1) \qquad\qquad (7.3\,b)$$

where s is the effective focal spot size and M is the projective magnifica-tion.

Then, taking the film unsharpness at 100 kV as 0.05 mm and consider-ing a microfocus X-ray tube with an effective focal spot size of 15 μm, there is no increase in the total unsharpness on the film until a value of $M = 4.3$ is reached, for $U_g = U_f$. For magnifications larger than this, there will be an increase in total image unsharpness. As the image is magnified, the true unsharpness related to the image detail size is U_T/M, which in this example is about 0.02 mm. This means that the graininess of the film is less obtrusive on image detail by this magnifica-tion factor, when compared with a contact radiograph. However, because X-ray outputs are small, exposure times may be long and it may be necessary to use a fast film/screen combination such as salt intensify-ing screens (typically $U_S = 0.3$ mm), and with this example a minimum magnification of $0.3/0.015 = 20$ would be needed to overcome the screen unsharpness.

With large values of projective magnification, the area of specimen examined at one time is small. To cover a reasonable area, large source-to-film distances have to be used, but this increases exposure-times and again necessitates the use of faster film/screen combinations. The second practical disadvantage of the method is that the film size required is the specimen size multiplied by the magnification, which makes it a very expensive technique in terms of film consumption.

The use of projective magnification techniques on film has so far been limited and a more widespread use of microfocus equipment has been with radioscopic systems, to minimize the imaging deficiencies of the primary conversion screen (Chapter 11). Projective magnification techniques have been used to examine high-alloy turbine blades, to separate porosity from diffraction spots. But if large magnifications are used, only very small areas of blade can be examined at each exposure.

Special non-radiographic film techniques

On many small assemblies there is a requirement for radiographs on which very high contrast is not necessary, but on which very fine detail rendition is essential. This requirement can be met by the use of single-emulsion films marketed for lithographic work. These films are generally known as 'Line' or 'Process' emulsions. Processed in conventional X-ray developers for slightly shorter than normal times, they yield a moderate contrast ($G_D = 2.5$ at density 2.0), a reasonable working density range (1–2.8), but most importantly because they have a very high silver-to-gelatin ratio, they have a small unsharpness. Used with 100 kV X-rays, a typical 'Line' type film has been measured to have an unsharpness of 0.02 mm compared with 0.05 mm for fine-grain film (0.02 mm is on the limit of what can be measured with the present techniques, and so must be treated with some caution).

For thin specimens placed close to the film, therefore, the total unsharpness can be very small, and also the resulting radiographs can be examined optically at magnifications up to × 30 before the graininess begins to interfere with image detail. For the examination of small assemblies, such as, for example, small detonators, a few hundred specimens can be radiographed on one film, instead of three or four at a time by projective magnification techniques. Normal practice would be to radiograph a single row at a time to maintain accurate alignment with the X-ray beam, and there would be up to 50 specimens per row.

The use of this type of film has several other advantages.

1. The full range of available X-ray kilovoltages can be used and is not limited to the very low range of microfocus equipment.
2. Films of this type have a maximum density of about 3.5 which means that the film never gets so black that one cannot see through it. This

is a considerable advantage when looking for the true edge of a cylindrical specimen, for example, for measurement purposes.
3. As conventional X-ray developer is used, variable development times are possible, and for the radiography of assemblies this has been varied from 1.5–5 min for particular applications.

The larger film manufacturers will usually supply their slowest radiographic film as single-sided coating. These have characteristics intermediate between the Line type emulsions and the conventional double-sided radiographic films. The contrast will be higher than on Line films, but as the silver-to-gelatin ratio is the same as on conventional film, there is little improvement in unsharpness.

Colour radiography

During 1965–70 there were several proposals for taking radiographs on colour film [8–10] and, by various tricks in processing, obtaining a radiograph exhibiting a wide range of colours. A colour radiograph of a complex specimen with a range of thicknesses will have an exceedingly complex image (part negative, part positive), with a range of densities and a range of most of the colours in the visible spectrum. Interest in the process seems to have disappeared and from experiments by the Author [10] several conclusions have been drawn.

1. Normal radiographic films give a much better sensitivity, over a limited thickness range, than colour film.
2. Line type film (developed for 2 min) can give as great a thickness latitude as colour film, using the same X-ray energy, with a markedly better sensitivity in the centre of the thickness range where the colour image changes from negative to positive, and approximately the same sensitivity at both extremes of the thickness range. Such film is also cheaper than X-ray film, and less than one-third of the cost of colour film. Its effective speed with X-rays is still about five times that of colour film.
3. The positive/negative dual nature of the colour film image means that there is one part of the density–thickness curve which has negative contrast, and the other has positive contrast; as the change from one region to the other is gradual and continuous there must be a region where the contrast is zero. In this region the image is seen (if it is seen at all) only by colour changes. The practical effect of this can be a region of poor discernibility of detail over part of the image of a complex specimen, and this poor region may obviously sometimes be a critical part, if the exposure time is misjudged.
4. Edges and gaps almost always show colour fringes. Some investigators have claimed that these enhance visibility, but in practice

spurious fringes usually complicate all but the simplest of images. The colour fringes at narrow gaps and around thin metal walls make it difficult to know what is the true thickness or spacing.

5. Slight variations in exposure or in processing, or in the light flash, can cause enormous changes in image colours and only the most rigorous control of all exposure and processing details can produce predictable and consistent results. These variations can lead to great difficulties in interpretation.

Fluorometallic intensifying screens

Although calcium tungstate (salt) intensifying screens are now rarely used in industrial radiography, since 1970 a special type of screen has been marketed by the Japanese Kyokko Company and more recently by Agfa-Gevaert. These screens consist of a thin layer of fluorescent material (calcium tungstate) coated on a thin lead foil and mounted on a support sheet. They can be used as conventional salt intensifying screens, but are mainly intended to be used with non-screen type film, when it is claimed that they permit a reduction of exposure time by a factor between five and ten with no loss of detail sensitivity compared with the same film used with lead intensifying screens. Kyokko screens are made in several coating thicknesses and are suitable for different radiation energies.

The calcium tungstate layer, which is between 100 and 200 µm thick, produces ultra-violet light while the lead layer produces electrons. All photographic emulsions are sensitive to electrons, but to obtain some increase in speed over lead screens there must be some additional sensitivity to ultra-violet light. If the sensitivity to ultra-violet light is too great, the result will be much like a salt screen exposure: a high intensifying factor, but a poor quality image. The choice of film is therefore very important, and only a very few non-screen films have the required ultra-violet response. On many commercial non-screen type films, the ultra-violet sensitivity is suppressed and such films are of no use with fluorometallic screens. Agfa-Gevaert [11] has produced a special film for use with these screens which has also been specially designed for rapid processing (total time 2 min at 30°C).

Experiments [12] have shown that the speed factor varies with the radiation energy and reduces as higher energy radiation is used. Experiments also showed that with any particular radiation the speed factor of fluorometallic screens varies with the actual exposure time.

Similar results were obtained with Ir-192 gamma-rays: the speed factor varied from 3:1 to 2:1. This effect is well known with salt intensifying screens, and is due to reciprocity law failure in the effect of the ultra-violet light on the film. It means that one important potential use of fluorometallic screens, to extend the maximum thickness which can be examined with any particular radiation, is not worthwhile.

Hiraki [13] showed that the speed ratio between fluorometallic screens and lead screens is temperature dependent. The relative speed of fluoro-metallic screens is greatest at − 60°C, when it may be four times greater than at 20°C. This point is of some practical interest in the radiography of arctic pipeline welds.

In summary:

1. when used with an appropriate film and X-rays between 80 and 200 kV, fluorometallic screens permit a reduction of exposure time up to about one-ninth of that required on lead screens, with no loss of IQI sensitivity;
2. alternatively, the X-ray energy can be reduced and the IQI sensitivity is slightly improved;
3. with 400 kV X-rays the reduction in exposure time is between 3:1 and 2:1;
4. with gamma-rays from Co-60 and Ir-192, the exposure time reduction is about 2:1.

The physical robustness of the screens has not been tested thoroughly but the base material is thin, high quality card, and they appear to be more robust than conventional salt screens. A screen has been bent to a 50 mm radius several times with no sign of damage.

These screens are widely used in Japan. It is understood that their chief application is in shipyard weld radiography to shorten exposure times and to enable low output, very lightweight X-ray sets to be used.

Fluorometallic screens are also used for the radiography of circum-ferential welds on lay-barge pipe laying. In this application there is a very restricted time-interval in which to perform any NDT before the newly-welded pipe is wrapped, slid into the sea and welding of the next joint begins. Very short exposure times with fluorometallic screens and high-speed film processing are used for very rapid radiography.

The use of very low kilovoltage X-rays

If X-ray kilovoltages below about 40 kV are used, some new factors need to be considered.

Almost certainly an X-ray tube with a thin beryllium window (1 mm or less) will be needed, to reduce the inherent tube filtration, and because most specimens will be thin, unsharpness becomes relatively unimport-ant; contrast and graininess become the major factors in image quality. Graininess is important because one assumes that discernible image detail is of the order of 1–4% of specimen thickness, so that this detail will be very small in all its dimensions.

In general there are no codes of good practice covering this field of radiography. The special points of technique to be covered are indicated below.

1. With metal (lead) intensifying screens, below about 120 kV there is very little intensifying action, but it is useful to retain the back screen if only to prevent backscatter reaching the film. A front screen (as thin as possible) acts as a clean-up filter at all X-ray kilovoltages, but below 60 kV requires an increase in exposure time and is better omitted. If the edges of a specimen are within the image area a very thin front screen should be retained down to at least 60 kV, but a conventional X-ray cassette front performs essentially the same function. Tin intensifying screens have been recommended and can be used advantageously at lower kilovoltages instead of lead screens.
2. Below about 50 kV the conventional cassette produces considerable extra absorption and should be replaced by one with a thin black polythene front (0.08 mm thick).
3. Below 20 kV the air between the X-ray source and the film begins to provide considerable absorption (see Table 7.8). The simplest technique to minimize this problem is to have a balloon of helium between the X-ray tube and the specimen, with the tube output window pressed into one side of the balloon. Alternative techniques, such as a helium-filled drum with foil end-faces, are also practical.

Figure 7.11 shows standardized exposure data for 20–80 kV X-rays for aluminium specimens. It is emphasised that these curves are standardized to 1 m source-to-film distance, which is not necessarily the recommended source-to-film distance for all specimen thicknesses in this range.

Dowd [14] studied the effects of kilovoltage on IQI sensitivity on aluminium specimens for a standardized technique using Kodak MX film, thin lead screens, density 2.0, and an exposure of 20 mA min. Bearing in mind that a shortening of source-to-film distance will allow a reduction in kilovoltage for a fixed exposure, he suggested two empirical formulas for the choice of optimum kilovoltage V and source-to-film distance D:

Table 7.8 Relative exposure ratios air/helium (280 mm)

X-ray kilovoltage	Exposure ratio air/helium
20	1.0
15	1.5
12	1.5
10	1.6
8	3.0
7	4.0

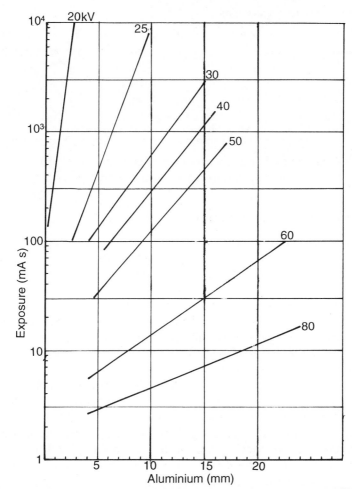

Fig. 7.11 Exposure curves for low energy X-rays: the curves for 20, 25 and 30 kV X-rays are for a beryllium window tube; 1 m focus-to-film distance, CX Film; density 2.0, polythene-fronted cassette; 4 min development at 20°C.

$$\left. \begin{array}{l} V = 3t + 33 \\ D = 39t + 400 \end{array} \right\} \tag{7.4}$$

t being the aluminium specimen thickness in millimetres.

The very marked effect of a change in kilovoltage can be seen in Table 7.9 for polythene specimens, with data measured from exposure curves obtained with a beryllium window X-ray tube.

There are no standard IQIs for very thin specimens. Very fine steel wires, down to 0.025 mm diameter (50 gauge), are easily obtained and can be built into an IQI which will measure unsharpness by acting as a

Table 7.9 Polythene specimens

Kilovoltage	μ (cm⁻¹)	Half-value thickness (cm)	Thickness for average exposure (cm)
7	18.2	0.038	0.8
8	14.5	0.048	1.0
10	10.0	0.069	1.5
12	7.3	0.094	2.0
15	5.0	0.137	3.0

high contrast detail. For use with plastics, very thin foils of some plastics are available and a simple step-wedge can be constructed.

Radiography with a moving beam

Radiography by scanning the specimen with a collimated X-ray beam, i.e. a narrow slit of X-rays, is possible and has been applied to a number of problems where it is desirable to maintain the X-ray beam at a fixed, chosen angle along the length of a long specimen. A typical application [15] is the examination of fuel elements from an atomic pile where it is desired to measure lengths of pellets by showing the interface between individual pellets. Lang [16] showed that under some conditions the method can be more rapid, and therefore more economical, than taking a series of single exposures.

The disadvantage of the technique is that movement unsharpness causes an additional unsharpness in the direction of the motion, and detail sensitivity is therefore never as good as on static radiographs. If the scanning slit is made narrower to reduce this unsharpness, the scanning speed must be slower as the exposure required is a direct function of the time any part of the slit is over each point on the film. Alternatively, the kilovoltage can be increased.

It is unlikely that this method will be useful if flaw sensitivity is a criterion of performance. Lang quoted for a 10 mm specimen, for the in-motion technique, 2.7% wire IQI sensitivity and 300 kV X-rays, compared with 1.4% and 100 kV X-rays for the static method. From the theory it would seem that crack sensitivity values for cracks parallel to the slit direction would show still greater differences. The chief application of the technique would seem to be where measurements of spacing etc. on the radiograph are required [17].

REFERENCES

[1] International Institute of Welding (1968) *IIS/IIW–309–68*.
[2] Feaver, M. (1977) *NDT Int.*, **10**(1), 13.
[3] International Institute of Welding, (1976) *Doc. V–586–76*.

[4] König, A. (1888) *Sitz. Akad. Wissensch. (Berlin)*, 917.

[5] Robinson, G. and Ward, J. M. (1979) *Conf. Weld. & Fabr. in the Nuclear Industry*, Paper No. 15, BNES.

[6] Pullen, D. and Hayward, P. (1979) *Brit. J. NDT*, **21**(4), 179.

[7] Ratcliffe, B. J. (1971) *Brit. J. NDT.*, **13**(2), 55.

[8] Beyer, N. S. (1961) *Argonne Nat. Lab. Rep.*, ANL 6515.

[9] Parish, R. W. and Pullen, D. (1964) D. *Brit. J. NDT*, **6**(4), 103.

[10] Halmshaw, R. (1970) *Int. J. NDT*, **1**, 295.

[11] Borcke, E. (1980) *NDT Australia*, **17**(8), 9.

[12] Halmshaw, R. (1972) *Brit. J. NDT*, **14**(2), 45.

[13] Hiraki, M. (1971) *Jap J. NDI*, **46**(2), 88.

[14] Dowd, M. (1971) *Brit. J. NDT*, **13**(5), 155.

[15] Huggins, B. (1981) *Brit J. NDT*, **23**(3), 119.

[16] Lang, G. (1964) *Proc. 4th Int. Conf. NDT*, Butterworths, London, p. 144.

[17] Halmshaw, R. (1964) *Proc. 4th Int. Conf. NDT*, Butterworths, London, p. 129.

8

Sensitivity performance

8.1 INTRODUCTION

Chapter 6 and 7 discussed in detail how radiographic techniques are developed, the effects of the various technique parameters and the image quality indicator (IQI) patterns which are available. It is now necessary to establish attainable IQI sensitivity values.

An IQI sensitivity is not the same as flaw sensitivity, but there is a complex relationship between the two; if IQI sensitivity is improved, flaw sensitivity will also improve (section 8.3).

IQIs have a limited usefulness in industrial radiography. They serve to check on technique details to a limited extent; they enable some judgement to be made of the quality of radiographs when technique details are not available; they can be used to compare techniques; and they give limited assistance in assessing the dimensions of the flaws which are revealed by radiography.

8.2 ATTAINABLE IQI VALUES

A number of basic assumptions are necessary.

1. The IQI is normally placed on the surface of the specimen remote from the film.
2. Film viewing conditions (section 7.4), such as transmitted luminance, masking and background luminance, are such that no limitation is imposed on seeing the detail which is present in the image.
3. The film reader has satisfactory near-distance eyesight, with correcting spectacles if necessary. The reader also has the use of a low-power magnifier, depending on the size of detail discernible in the film image.

In considering IQI data, one must never forget that the results are subjective. If a number of experienced readers examine a radiograph under good viewing conditions it is unlikely that all will claim to discern the same number of wires or holes on the IQI image. Skilled film readers, however, are unlikely to vary by more than one IQI element (wire or hole). If the intervals in size between successive elements of the IQI are large, the spread over a set of observations will be very small: if the interval between elements is small, the spread will be greater, and this is one important factor in IQI design.

The different observers may have different criteria of the limit of discernibility, and this also causes a spread in the results when one film is examined by different observers. Some attempts have been made to set criteria for discernibility. In the step/hole IQI a reasonable criterion is that both holes in the steps having two holes must be detected. In the case of the wire IQI two criteria are possible:

1. to take the last wire, the whole length of which can be discerned;
2. to take the last wire of which there is any evidence.

Either criterion is justifiable, but the Author prefers and works to the second. If the specimen thickness under the IQI is not uniform, the first criterion is difficult to apply.

Finally, if a particular IQI is specified, it must be remembered that the attainable sensitivities for any one specimen thickness are a series of discrete values and not a continuous curve. If an IQI has wires of (say) 0.5, 0.63, 0.8, 1.0, and 1.25 mm, and is used on a 50 mm specimen, the only possible IQI sensitivity values are 1.0, 1.26, 1.6, 2.0 and 2.5%, whereas a table of attainable values may say that 1.5% is attainable. This may be the true limiting value, based on a smoothed curve, which would be obtained with a particular technique if there was a suitable element on the IQI, but in this particular example the film reader would have to record 1.6% as the best attainable.

Thus, in Fig. 8.1, the smooth curve represents the best attainable sensitivities for a range of steel thicknesses, using drilled-hole type IQI (hole diameter = step thickness). In terms of the actual IQI available, however, the saw-tooth line represents the best attainable values, as judged with the standard IQI sizes. To be of any use for assessing the quality of any radiographic technique, it is necessary to know what IQI sensitivities should be obtained with various techniques and on a range of thicknesses and materials. Unfortunately the change of IQI values with change in technique is not very large with existing patterns of IQI, so that it is necessary that any quoted acceptable values should be reasonably accurate; the agreement of such values has taken many years of patient experiment and discussion in several countries.

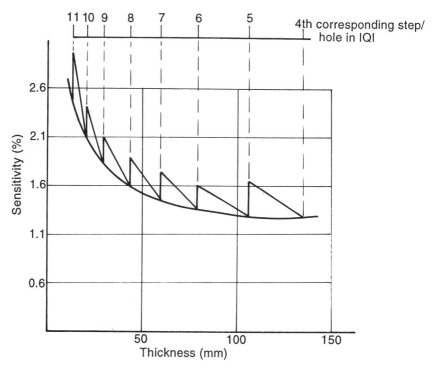

Fig. 8.1 Sensitivity values with a step/hole IQI. The smooth curve shows the attainable values, assuming a suitable step; the saw-tooth line represents the actual values obtainable with the standard IQI model.

Various standards (BS: 3971:1980, EN-1435: 1995 etc.) have proposed tables of IQI values which should be obtained if the techniques are correctly applied. Generally these are considerably poorer than the values that the Author has found to be possible (Table 8.1). Experiments have shown that almost exactly the same values as are given in Table 8.1 for X-rays and steel can be obtained on the same thicknesses of aluminium and aluminium alloy specimens if the appropriate kilovoltages are used.

The main points which need to be emphasized on IQI are as follows.

1. With existing patterns, the changes in attained IQI sensitivity with changes in technique are not very large.
2. It is essential to relate the required IQI sensitivity to both the type of IQI and the specimen thickness. There is no pattern of IQI which produces 2% or 1% sensitivity over a wide range of specimen thicknesses, using analogous techniques.
3. Judgement of image discernibility is not a precise parameter, and some spread of IQI readings must be expected.

Table 8.1 Attainable IQI sensitivity values in steel

Specimen thickness (mm)	Source kilovoltage	Wire IQI (%)	Wire IQI diameter (mm)	Step/Hole IQI (%)	Step/Hole IQI Hole diameter (mm)	Step wedge (%)	Duplex wire (mm)
X-rays							
3	80	1.6	0.05	–	–	–	–
6	120	1.1	0.063	2.7	0.16	–	–
12	150	1.0	0.125	2.5	0.32	–	< 0.10
25	220	1.0	0.25	2.0	0.50	< 0.5	0.16
40	280	0.8	0.32	1.6	0.63	0.5	0.2
50	300	0.8	0.40	1.6	0.80	0.7	0.3
75	400	0.7	0.50	1.3	1.00	0.8	0.4
100	1 MV	0.6	0.63	1.6	1.60	0.5	0.4
200	8 MV	0.5	1.00	–	–	0.4	0.6
Gamma-rays							
3	Yb-169	2.7	0.08	–	–	–	–
6	Yb-169	2.0	0.125	5.0	0.32	–	–
6	Ir-192	2.6	0.16	6.6	0.4	–	–
12	Yb-169	1.3	0.16	3.3	0.4	–	–
12	Ir-192	1.7	0.20	4.1	0.5	–	0.10
25	Ir-192	1.3	0.32	2.6	0.63	0.6	0.10
40	Ir-192	1.0	0.40	2.0	0.8	0.55	0.13
50	Ir-192	1.0	0.50	2.0	1.0	0.5	0.20
75	Co-60	0.85	0.63	1.7	1.25	0.6	0.25
100	Co-60	0.8	0.80	1.6	1.6	–	0.40
150	Co-60	0.8	1.25	1.7	2.5	–	0.50

8.3 CALCULATION OF DETAIL SENSITIVITY

The basic equation for thickness sensitivity was established in Chapter 2 (equation 2.18). If Δx is the minimum thickness change which can be detected, ΔD must then be the minimum density difference detectable by eye on the radiograph, and $\Delta x/x$ can be expressed as a percentage thickness sensitivity S:

$$S = \frac{\Delta x}{x} \times 100 = \frac{2.3\,\Delta D}{\mu G_D x}\left(1 + \frac{I_S}{I_D}\right) \times 100$$

which can conveniently be written as

$$S = \frac{A}{x}\,100 \tag{8.1}$$

where

$$A = \frac{2.3\,\Delta D}{\mu\,G_D}\left(1 + \frac{I_S}{I_D}\right)$$

There is ample evidence that this equation agrees well with experiment. ΔD has a value between 0.006 and 0.01 depending on the viewing conditions. This formula can be extended to the calculation of the minimum size of drill hole or wire which may be detected on a radiograph and to such flaws as artificial cracks. The most difficult additional parameters which need to be incorporated are:

1. total radiographic unsharpness U_T;
2. the performance of the human eye when viewing very low contrast images of different shapes and sizes.

One method of developing formulas for detail sensitivity from equation (8.1) is as described below. Considering a very small cavity of width W and depth b in a specimen of total thickness t, the excess radiation penetrating the cavity is

$$I_0 W \left[\exp\{- \mu(t - b)\} - \exp(-\mu t) \right]$$

which approximates to

$$I_0 \mu (Wb) \exp(-\mu t)$$

Thus for a small wire of radius r the change in quantity of direct radiation reaching the film per unit length of wire is

$$\pi r^2 \mu I_D \qquad\qquad (8.2)$$

In the presence of image unsharpness, the image of a wire is broadened and reduced in contrast, and the width of the image on film is

$$(2r + U_T)$$

where U_T is the effective total unsharpness. As the characteristic of an X-ray film can be expressed by

$$D = G_D \log_{10} (I_S + I_D)T + \text{constant}$$

where T is the exposure, then

$$\Delta D = 0.43 G_D \frac{\Delta I_D}{I_S + I_D} \qquad\qquad (8.3)$$

If then it is assumed that the intensity of radiation across the image of the wire is constant, the change in radiation intensity in the image will be

$$\pi r^2 \mu I_D \left(\frac{1}{2r + U_T} \right) = \Delta I_D = \frac{\Delta D (I_S + I_D)}{0.43 G_D} \qquad\qquad (8.4)$$

Since the film density in the image is not uniform, it has been suggested that the normal minimum discernible density difference ΔD will not be the same as for a simple density-step, and a 'form factor' should be introduced, (F ΔD) instead of ΔD. For small wires $F = 0.6$ and for larger wires $F = 1$. The resulting equation is

$$\frac{r^2}{2r + U} = \frac{2.3F\,\Delta D\,(1 + I_S/I_D)}{G_D\,\mu\pi} \tag{8.5}$$

From this equation r can be evaluated and the wire sensitivity is

$$\left(\frac{2r}{t} \times 100\right)(\%)$$

A similar equation can be developed for step/hole IQI sensitivity and if the hole diameter equals the step thickness h then

$$\frac{h^3}{(h + 2U_T)^2} = F\,\frac{2.3\Delta D}{\mu G_D}\left(1 + \frac{I_S}{I_D}\right) \tag{8.6}$$

Although these equations must have limitations in that they take no account of film graininess, they give a good insight into the factors affecting IQI sensitivity. Values calculated from these equations are shown in Table 8.2 and can be compared with the experimental values given in Table 8.1.

From Table 8.2 it can be seen that the calculated wire IQI values agree well with experiment, and this is possibly because the wire is several millimetres long and not therefore much affected by the graininess of the image. The step/hole IQI values are a little more erratic, and in general the calculated values are slightly lower than experiment, indicating that graininess has more influence on a small circular image, as would be expected. Also, with very small images, the value to be taken for ΔD is less certain and probably varies slightly with image size; on a 6 mm specimen, a 2% hole has a diameter of only 0.12 mm.

Table 8.2 Calculated values of IQI sensitivity

Steel thickness (mm)	Calculated sensitivity (%)		Remarks
	Wire	Step/hole	
3	1.6	–	60 kV X-rays
6	0.9	2.3	100 kV X-rays
12	1.0	1.7	150 kV X-rays
12	1.7	3.4	Ir-192 gamma-rays
25	0.8	1.5	220 kV X-rays
25	1.1	2.1	Ir-192 gamma-rays
50	0.6	1.6	300 kV X-rays
50	0.95	1.6	Ir-192 gamma-rays
75	0.7	1.2	400 kV X-rays
75	0.74	1.5	Co-60 gamma-rays
100	0.6	1.3	I MV X-rays
100	0.7	1.9	Co-60 gamma-rays
150	0.55	1.6	5 MV X-rays

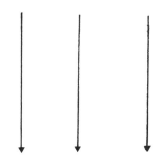

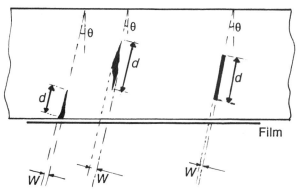

Fig. 8.2 Parameters of a surface-breaking crack, an internal crack and an artificial slit.

Crack sensitivity

A similar calculation can be used to develop a formula for crack sensitivity. Taking the crack parameters shown in Fig. 8.2, it can be shown that

$$dW = \frac{2.3 \Delta D\,F}{G_D \mu} \, (d\sin\theta + W\cos\theta + U_T)\left(1 + \frac{I_S}{I_D}\right) \qquad (8.7\,a)$$

or

$$dW = AF\,(d\sin\theta + W\cos\theta + U_T) \qquad (8.7\,b)$$

then

$$d = AF\left(\frac{W\cos\theta + U_T}{W - AF\sin\theta}\right) \qquad (8.8)$$

These equations show that the relationship between wire and crack sensitivity is not a direct one. Thus, two techniques giving the same wire IQI sensitivity do not necessarily give the same crack sensitivity. This can be confirmed experimentally by varying the geometric unsharpness;

the wire IQI sensitivity changes only very slightly, whereas the crack sensitivity varies much more.

To specify a 'crack' for the determination of sensitivity from equation (8.7) it is necessary to assign a width W to the crack, so that the cross-section is more that of a narrow slit than a natural crack which would taper in width. Since a natural crack tapers, the maximum width cannot be taken as the correct value for W in equations (8.7) and (8.8); a more realistic value would seem to be $W/2$.

Data on natural cracks are extremely scarce because of the difficulty in quantifying the crack dimensions, but comparison of equation (8.7) with experimental data on narrow slits shows good agreement, particularly with high energy radiation where unsharpness tends to be more important than film graininess.

8.4 FLAW SENSITIVITY

Although there is no simple relationship between IQI sensitivity and flaw sensitivity, and it certainly is not true that if the IQI sensitivity is 2% this means that all flaws occupying 2% of the thickness are detectable, it is reasonable to suggest that some direct relationship must exist. There are four types of commonly-occurring flaws where the use of an IQI should give a reasonable estimate of flaw sensitivity. In each case the flaw is air-filled; if it is composed of slag, dross or sand etc., the relationship with IQI detail is invalid. The four types of flaw are described below.

1. Small pores or gas holes. These occur in welds and castings and are usually spherical in shape. As the relative volumes of a spherical cavity and a cylindrical hole in a step/hole IQI both with radius r is 2:3, the step/hole IQI sensitivity gives a good approximation with this correcting factor. Thus, if an IQI sensitivity of 2% is obtained, the porosity sensitivity should be about 3%.
2. Piping. This is a linear discontinuity of roughly circular cross-section occurring in castings as shrinkage piping. If it is of small diameter the piping corresponds directly to a wire, except that in one case material has been added to the thickness and in the other subtracted. Wire IQI sensitivity therefore directly measures 'piping' sensitivity.
3. Defects with a considerable area (e.g. large voids in a casting). If the defects have an area considerably larger than the acting unsharpness, sensitivity is largely a matter of contrast and corresponds to simple thickness-sensitivity as measured with a step-wedge.
4. Cracks. If these are very narrow, with a large depth/width ratio, there is no standard IQI with this type of detail, but it has been shown that there is a relationship between crack and wire IQI sensitivity. If the total acting unsharpness is known, one may be calculated from a

knowledge of the other. The same relationship holds for other very narrow flaws, such as lack of root penetration.

Effect of flaw position

The position of the flaw through the thickness of the specimen has some effect on sensitivity. As the geometric unsharpness depends on the distance of the flaw from the film, it becomes progressively smaller for flaws closer to the film, and if the flaw is narrow, such as a crack, there is an improvement in sensitivity for flaws near the film. In the case of a broader flaw this effect is less marked. Due to the effects of scattered radiation a flaw on the film surface of the specimen, or very close to it, is shown with a slightly enhanced contrast. If a thick front intensifying screen is being used this latter effect is virtually eliminated.

8.5 DEFECT DEPTH DETERMINATION [1]

Most defects found by radiography are three-dimensional and from the point-of-view of deciding the significance of a defect it is desirable to be able to determine all three dimensions, as well as the nature of the defect. Generally, by radiography, one can recognize the nature of a defect and can easily measure its effective length and width parallel to the plane of the film, but the through-thickness dimension (height) is less easy to determine. The distance of a defect from the surface (depth) can be found by stereometric methods. Some information on the through-thickness dimension of a flaw can be obtained from a densitometric scan across the image, but this does not work well with narrow flaws such as cracks.

An alternative technique was proposed by Yokota and Ishii [2], the essence of their method being to take two exposures at different angles, (typically $\pm 10°$) and measure the width of the two crack images. From these two measures the crack height can easily be calculated, irrespective of whether the crack is normal to the surface or at an angle. If the two angles of the X-ray beam are $\pm \theta°$, and the crack images have widths of l_1 and l_2, the height of the crack h is

$$h = \tfrac{1}{2}(l_1 + l_2)\cot\theta \qquad (8.9)$$

Trials with different methods of image measurement have shown that a high-power film projector, with a linear magnification of $\times 15$–20 and good quality optics, gives the most consistent results. With this method, crack image widths in the range 0.2–0.8 mm could be measured to an accuracy estimated at $\pm 10\%$, which appears to be similar to the accuracy claimed by the Japanese workers. Yokota and Ishii [2] claimed that it is desirable to keep the total unsharpness below 0.2 mm, which means that they prefered to have the crack on the film side of the specimen; they also suggested that extra-fine-grain film be used. They claimed that, overall, the height of natural cracks was consistently underestimated by about 25%, due to tightness of the propagating edge of the crack; their cracks had opening widths of 0.06–0.2 mm. With high energy X-rays, in plates 60–100 mm thick, Morikawa and Koyama [3] claimed errors not greater than ± 4% in determined crack height. These proposals are simple and depend entirely on the ability to measure accurately the width of an image of a defect when the radiograph is deliberately taken at an angle. Experiments have shown that this is possible, although difficult, even on a crack image.

8.6 RADIOGRAPHIC IMAGING CONSIDERED IN TERMS OF SPATIAL FREQUENCIES

The treatment of radiographic imaging used so far, as a problem of contrast, definition measured as a series of unsharpnesses, and film granularity, can take the problems of industrial radiographic images quite a long way towards a solution. However, the problems of granularity (or image noise) and those of non-uniform focal spots have been side-stepped rather than solved, and newer ideas from the field of photographic imaging have shown that a completely different approach to the problem is possible.

A complete radiographic system from radiation source to detector can be regarded as an image transfer process using an imperfect recording system. In the transfer process the image is modified and in the recording system there are further limitations:

$$\text{Signal}_{\text{recorded}} = \text{Signal}_{\text{in}} \, [M\,(f)]^2 \, [T\,(f)]^2 \qquad (8.10)$$

where $M(f)$ represents the transfer characteristics of the radiographic system, i.e. the geometric unsharpness, and $T(f)$ represents the recording characteristics of the films and screens, i.e. the film unsharpness, screen unsharpness etc., both expressed in terms of spatial frequency f in lines per millimetre. The transfer functions are included as squared values as will be explained later, because signals are usually expressed as intensity functions whereas modulation transfer functions (MTFs) are amplitude functions.

From Fourier analysis any input signal to a system can be expressed as a series of sinusoidal curves of different amplitudes and frequencies, so that a complete frequency-response curve of the transfer characteristics should specify the performance of a system in transmitting any chosen input signal.

A perfect transfer curve would be as shown in Fig. 8.3(a), whereas a practical transfer characteristic would be more like Fig. 8.3(b), which esentially says that the spatial frequencies from 5 to 20 lines mm^{-1} are

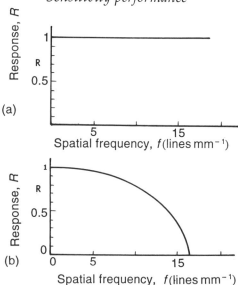

Fig. 8.3 Spatial frequency transfer curves. (a) = perfect transfer characteristics; (b) = practical transfer function.

transmitted through the system with reducing intensities, and frequencies above 20 lines mm^{-1} are not transmitted at all.

The building bricks of this theory are the line spread function (LSF), the MTF and the concept of spatial frequency, e.g. lines per millimetre.

If an X-ray beam is passed through a very narrow metal slit and the resulting 'line' image is recorded on film, the film density distribution will be as shown in Fig. 8.4. This is the LSF of the set-up used, characterizing the parameters controlling image definition.

If a test object is taken, consisting of a series of regularly-spaced bars and spaces (the bars being assumed to absorb the X-rays), and a radiograph made, a perfect X-ray system should produce a spatial density distribution similar to that shown in curve A in Fig. 8.5. It is probable, however, that if a finely-spaced bar pattern is used, that the density distribution will be more like that shown in curve B in Fig. 8.5, where the edges of the pattern are slightly blurred and the density difference between the images of the bars and spaces is reduced compared with curve A. If a series of bar/space patterns of different spatial frequencies is taken, a point will be reached where the system is incapable of resolving the pattern. The density difference between bar and space in the image is called the image contrast or response R (or modulation), and the fineness of the pattern is the spatial frequency. A curve of R against f is defined as the MTF if only the modulus of R is taken. If the frequency response curve is plotted taking account of phase effects, i.e. negative values of R where necessary, the curve is referred to as an optical transfer

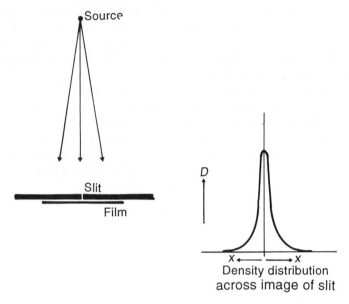

Fig. 8.4 Line spread function of a radiographic film.

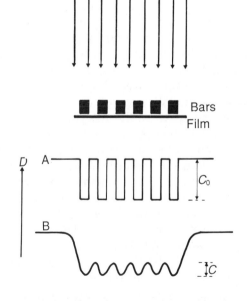

Fig. 8.5 Modulation transfer function with X-rays.

function (OTF). These negative values are of considerable importance when practical non-uniform radiation sources are considered.

Strictly, the intensity distribution produced by the bar/space test pattern should be sinusoidal rather than square-wave. If film densities are used rather than radiation intensities, it must be assumed that there is a linear conversion from intensity to density. The response, assuming no image deterioration, is usually normalized to unity, i.e. $R_{max} = 1$.

In optical work the R/f values can be found directly by using a series of bar/space patterns of different frequencies as input, but with X-rays it is more practical to determine the MTF or OTF curves mathematically from experimental LSF curves by convolution with a sinusoidal function. If the LSF is $S(x)$, and is symmetrical, then

$$R(f) = \frac{\int S(x) \cos 2\pi fx}{\int S(x) \, dx} \, dx \qquad (8.11)$$

Spread functions in space can be convoluted, so that a spread function due to geometrical unsharpness $f(x, y)$ is combined with a film with a spread function $g(x, y)$ to give a joint spread function $h(x, y)$:

$$h(x, y) = g(x, y) * f(x, y) \qquad (8.12)$$

The corresponding transfer functions F, G and H in terms of spatial frequencies u and v can be simply multiplied:

$$H(u, v) = G(u, v) \times F(u, v) \qquad (8.13)$$

F, G and H being the respective Fourier transforms of f, g and h.

In non-mathematical terms, if an elemental bit of a primary image is taken, a perfect transfer system would reproduce this as the same amplitude and width in the final image, whereas in an imperfect system each element is spread out in width and possibly also reduced in amplitude, the final image being the sum of all these elemental images.

There is a close relationship between LSF and film unsharpness. It can easily be shown that a plot of the slope of the unsharpness curve against x gives the LSF curve. Conversely, if the area under the LSF curve is determined for different values of x the unsharpness curve is obtained. This concept of transfer function can be used to express the effect of each stage in an imaging process, because it has been shown that MTF values can be multiplied. By using Fourier transform methods, any spatial intensity distribution of X-rays or light, with distance as the variable, can be transformed into a function with spatial frequency as the new variable. Thus, the intensity distribution in space of the image of a small detail can be expressed as a function of spatial frequency and the effect of an imaging system of known MTF on the image can be calculated. The advantages of using MTF or OTF to measure image degradation are as follows.

1. The result is completely objective and is not dependent on the judgement of an experimenter.

2. MTF data can be obtained for each stage of an imaging system and then cascaded to produce a curve for the complete system. This enables one to study individual stages and decide which ones are controlling the overall performance. This is especially valuable in studies such as television–fluoroscopic systems (Chapter 11).
3. Unsharpness and resolution data can be derived from MTF data.

The weakness of MTF data is that they do not take account of the performance of the eye in respect of varying image shape. Lamar and Hecht [4] showed that the minimum discernible contrast of images of the same area is a function of the length-to-width ratio of the images.

Experimental determination of MTF values

If an experimental LSF curve can be produced with sufficient accuracy, MTF values can be obtained. They can be presented either as a continuous curve or as a table of R/f values (Fig. 8.6 and Table 8.3). The MTF values of films 1, 2, 3, 5, 6, 8 are very similar, within experimental error, although these films cover a speed range of about 5:1 and are from three different manufacturers. Films 4 and 7 were very high-speed films, but are no longer available; they clearly have much different MTF curves. Film 9 is a very slow non-radiographic film for comparison.

By determining the intensity/density distribution across a pinhole image of the focal spot of an X-ray tube the OTF curves corresponding to

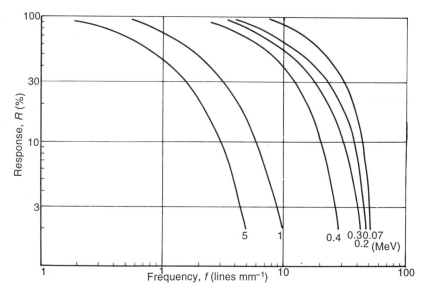

Fig. 8.6 Modulation transfer function curves determined for one radiographic film exposed to different X-ray energies (from 70 kV to 5 MV).

Table 8.3 Experimental values for MTF for different radiographic films (250 kV X-rays with 0.3 mm lead filter; 0.05 mm lead intensifying screens)

Film	\(f \) (lines/mm)								
	2	5	10	15	20	25	30	40	50
1. Agfa-Gevaert D.2	97	82	60	42	28	17	10	3	–
2. Agfa-Gevaert D.4	98	90	66	49	33	21	13	3	–
3. Agfa-Gevaert D.7	98	88	65	46	31	20	12	2	–
4. Agfa-Gevaert D.10	97	81	46	22	9	1	–	–	–
5. Kodak MX	97	84	60	43	30	19	10	–	–
6. Kodak CX	98	85	65	49	30	18	7	–	–
7. Ilford G	95	71	39	16	6	1	–	–	–
8. Ilford B	95	81	54	35	19	14	8	2	–
9. Ilford N.5.50 single emulsion	98	90	70	55	47	36	21	8	2

the focal spot can be calculated (Fig. 8.7). As the focal spot output is non-uniform the OTF values correspond only to one line across the focal spot, and have to be calculated for each different scan direction. The practical values differ considerably from values calculated for a theoretical uniform spot.

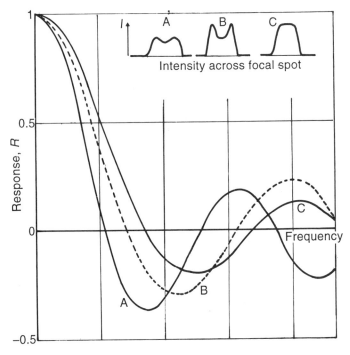

Fig. 8.7 Transfer function curves for the focal spot of an X-ray tube, determined for three different directions across the focal spot.

Mizunuma [5] suggested that the MTF curve obtained from the magnified image of a wire can be used to determine the practical effective width of a non-uniformly emitting focal spot. Comparison of his results with other methods of focal spot measurement show reasonable agreement.

Uses of MTF measurements

The main use of MTF measurements in radiography is to study the interplay of factors in a radiographic set-up, such as the effect of focal spot size. Although MTF measurements can provide much more quantitative data about the performance of a radiological system than any previous method of assessing performance, they still do not provide a complete answer to the specification of detail-recording ability, because they do not take into account the grainy nature of the photographic image nor the performance of the human eye on the final perceived image. In addition, some radiological systems may be limited by the number of utilized quanta (see discussion below).

Wiener spectrum of granularity

For the classification of radiographic films [6], the standard deviation σ_D of the density readings produced when a film of uniform density D is scanned with a light spot is used as a measure of granularity. This is 'Selwyn granularity', but as long ago as 1955 Jones [7] suggested that Selwyn granularity was not adequate to fully describe film granularity and that a more comprehensive description could be given by a film noise spectrum, now generally known as the Wiener spectrum of film grain. The Wiener spectrum has the same relation to granularity as does the power spectrum of electric circuit theory to electrical noise. It uses the same parameter (spatial frequency) to describe granularity, as is used for the MTF.

The irregular distribution of density of a uniformly exposed sample can be analysed by Fourier methods into sinusoidal components; if the sample is large, the range of frequencies is also large and may be thought of as a continuous spectrum which is equivalent to the Fourier transform of the density distribution. Although, therefore, the standard deviation σ_D and the Wiener spectrum contain intrinsically the same information, there is a definite advantage in using the Wiener spectrum if it is desired to use it to study granularity as the 'noise' of the image, as noise in this context is directly additive.

The standard deviation σ_D is the square root of the area under the curve of the Wiener spectrum:

$$(\sigma_D)^2 = \int W(f)\, df \tag{8.14}$$

The accuracy of the determination of the Wiener spectrum has been the subject of many papers, notably by de Belder [8] and Shaw [9] and there is considerable disagreement on both the magnitude and shape of Wiener spectrum curves, particularly those for films exposed to X-rays.

Experimentally, a high-precision microdensitometer is needed, with sufficient axial depth-of-field in the imaging system to cover a double-emulsion radiographic film. The Author [10] used a scanning spot of $200 \times 2.5\,\mu m$ and a minimum of 7000 readings to calculate each spectrum. The area of the film scanned was kept to a minimum, to reduce any effects due to local density variations across the film. A few of the curves obtained are shown in Figs 8.8 and 8.9, and in general the results and those of other workers [11] appear to justify four conclusions.

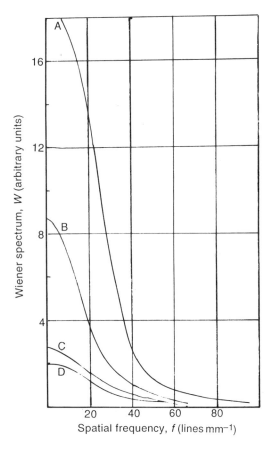

Fig. 8.8 Wiener spectra for different radiographic films exposed to 400 kV X-rays: all films exposed with lead intensifying screens. Curve A = very fast film; Curve B = medium-speed film; Curve C = fine-grain film; Curve D = very-fine-grain film.

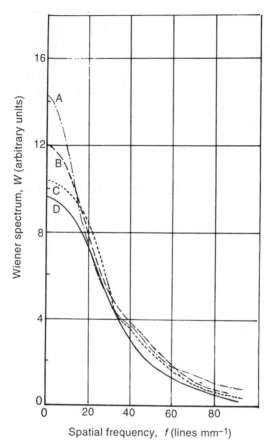

Fig. 8.9 Wiener spectra for a medium-speed radiographic film exposed to different X-ray energies. Curve A = 31 MV X-rays; Curve B = 400 kV X-rays; Curve C = 100 kV X-rays; Curve D = 50 kV X-rays.

1. For films exposed to X-rays, even using salt intensifying screens, the spectrum is not white noise, but varies in intensity with frequency.
2. The intensity of the spectra, particularly at low frequencies, varies markedly with the type of film.
3. The same type of film exposed to different energies of radiation shows only slight differences in Wiener spectrum values at high frequencies above $f = 20$ lines mm^{-1}, but considerable differences occur at low frequencies (Fig. 8.9).
4. The value of σ_D increases rapidly as one changes to film of higher speed, but the ratio of values is not as great as the film speed ratio.

The third conclusion appears to agree closely with direct observation of film graininess when the same film type is exposed to different radiations. Careful visual observation suggests that there is a uniform pattern of fine grain with an overlying, much coarser pattern of 'strings' of grains which become apparent with X-ray energies above about 400 keV, and gradually become more dominant as the X-ray energy increases. This coarse pattern corresponds to the low-frequency components of the Wiener spectrum and can be regarded as the first visual evidence of quantum fluctuations (quantum mottle) imposed on the intrinsic granularity of the film [10].

Determination of a film quality index

It is possible to combine film contrast, MTF and Wiener spectrum to form a specification of film performance. Qualitatively if film contrast is increased a given size of object detail is recorded as a larger density difference, so that an increase in contrast reduces the minimum size of object detail which can be detected, if other factors are constant or do not provide an overriding limitation.

An input signal can be considered two-dimensionally as being represented by an intensity (power) spectrum $P(f)$, where intensity is plotted against frequency f. For the moment, quantum fluctuations in the input signal will be ignored. This input then passes through a system which has a filtering effect represented by its modulation transfer function $T(f)$. Finally the signal is recorded by a film which also has an MTF $R(f)$, a Wiener spectrum $W(f)$ and a contrast G. MTFs are amplitude/frequency functions so they must be squared to bring them to the same dimensions as intensity spectra. Thus the signal, as recorded on film, can be represented by

$$S = P(f) [T(f)]^2 [GR(f)]^2 \tag{8.15}$$

and the signal-to-noise ratio can be written as

$$\frac{S}{N} = \underset{\underset{\text{signal}}{\downarrow}}{P(f)} \quad \underset{\underset{\text{system}}{\downarrow}}{[T(f)]^2} \underset{\underset{\text{film}}{\downarrow}}{\frac{G^2 [R(f)]^2}{W(f)}} \tag{8.16}$$

If $P(f) [T(f)]^2$ can be regarded as constant over the range of frequencies to be considered, this term can be taken outside the integration, and the remainder, integrated over all values of f, can be taken as a 'film quality index' F (FQI):

$$F = G^2 \int \frac{[R(f)]^2}{W(f)} \, df \tag{8.17}$$

Some workers consider that for such an FQI to be practically useful, it should be further modified by the MTF of the viewing eye determined for the conditions of viewing the image.

In theory one could take an image of a flaw, express this in spatial frequency terms and determine the effect of the characteristics of a specific film on the image. Similarly the effect of an imaging system such as radioscopic equipment could be determined. Putting in experimental values gives FQI values from 11 000 for very fine-grain film used with low X-ray energies, to 40 for fast film used with megavoltage X-rays [12], if the full frequency range of the films is taken. Different films used with 100 kV X-rays show a range of 60:1, which is almost the same as the speed range of the same films. Using one type of film, between 50 kV X-rays and 1 MV X-rays there is a FQI reduction of 42:1.

Quantum fluctuations

So far, contrast has been treated in a simple, straightforward manner as a density difference on an illuminated radiograph, or a luminance difference between an image and the background on a fluorescent screen image. Obviously there must be a minimum value of contrast which the eye is capable of detecting, ΔD. Generally, in industrial radiography, one thinks in terms of ΔD being measured across the image of an edge but there is considerable experimental evidence to show that ΔD is not constant, but is dependent on the shape and size of the image detail.

There is no general formula relating image contrast to image shape and size to discernibility. There is, however, another way of considering contrast, which is of great importance in radiology and has also been applied to more general image perceptibility problems.

8.7 NOISE LIMITATIONS

The image on a radiograph (and on a fluorescent screen) is formed by the effects of a beam of X-ray quanta, whether these are directly incident on the detector or are converted into light or electrons before absorption and recording. The emission of X-ray quanta from the target of an X-ray tube and the absorption processes in the specimen and in a fluorescent screen are random phenomena, that is, while one can give an average to the number of quanta involved in forming unit area of image, this is not an exact number. There is a natural fluctuation in the number of quanta involved, and statistical analysis shows that if N quanta are of concern the average fluctuations is $N^{1/2}$. If N quanta are absorbed in unit area of a film in an exposure time t, then when

$N = 100$ quanta, the average fluctuation is $N^{1/2}$; i.e. 10% of N

and when

$N = 10\,000$ quanta, the average fluctuation is $N^{1/2}$; i.e. 1% of N.

If, therefore, one considers a small circular image on a screen, which is slightly different in luminance from the background, and this area is formed by the absorption of, on average, 100 quanta, there will be an average fluctuation in luminance of this small image of 10%, due to statistical fluctuations in the number of quanta utilized in forming the image. If the image is only, say, 5% brighter than the background, on average, the observer will not see the image because the fluctuations will mask the real difference. If, however, the same small image is formed as the result of the absorption of 10^4 quanta, the statistical fluctuations will be only 1%, and the 5% real difference postulated should be observable. There is strong evidence that for a real difference in contrast to be observable, it must be at least three times the statistical fluctuation level; some workers have suggested a factor of five.

A 5% luminance difference is a film density difference of 0.02, which is well within the capabilities of the eye, under good film viewing conditions. It is extremely important to realize that although there may be a very large amplification factor in the system, such as in an X-ray image intensifier or a CCTV system, so that the number of light quanta reaching the eye is many times larger than the number of X-ray quanta forming the image, the percentage fluctuation at the stage in the imaging process when there is the smallest number of utilized quanta (which is nearly always that at the primary conversion screen) is transmitted through the system. This was very clearly brought out in a classic paper on X-ray fluoroscopic systems by Sturm and Morgan [13].

The argument makes no assumptions whatever about screen or film granularity.

To detect small contrast differences, therefore, it is essential to utilize as many quanta as possible on the primary screen and the minimum contrast observable, irrespective of any instrumental deficiencies, will depend directly on the number of quanta utilized.

In film radiography exactly the same argument holds except that here the number of quanta are integrated over the exposure time. A slow film, requiring a longer exposure time to produce a standard density, will enable a lower minimum contrast to be detected.

It is important to emphasize that signal-to-noise values must be related to image area (or to measuring aperture) and that scattered radiation degrades contrast by $(1 + I_S/I_D)^{-1}$ where I_S and I_D are the scattered and direct components of the radiation reaching the detector.

It has been suggested by some workers, notably Fellgett [14] working with light and de Belder [15] with X-rays, that film granularity might be fully explained as a manifestation of photon noise in the incident radiation. Other workers [16] have rejected this argument, but the controversy appears to have subsided in recent years.

Rose [17, 18], in a series of papers from 1946 onwards, attempted to use the concepts of quantum fluctuations and signal-to-noise ratio to specify the performance of all imaging pick-up devices, e.g. film, television tubes, image intensifiers and the human eye.

It may be thought that this concept of spatial frequency as a parameter makes the imaging problem with X-rays unnecessarily complicated, but the concepts of the three treatments outlined are closely related. Fundamentally, an image can be studied and defined in terms of contrast, definition and noise, noise being taken to cover quantum fluctuations, granularity, screen mottle etc. The simplest way to deal with image definition is in terms of unsharpness, but in the practical world of non-uniformly emitting sources of radiation, this treatment becomes too simplified and there are considerable advantages in using MTF instead. An additional advantage of using spatial frequency as a variable is that film granularity, or quantum mottle as exhibited on a film or screen, can also be expressed in the same terms, i.e. the Wiener spectrum of the (noisy) image.

Quantum fluctuation calculations can be used separately to relate detectable image size to minimum detectable contrast.

X-ray imaging is a special case in that the number of X-ray quanta utilized in the primary conversion screen is almost always the limiting number controlling the quantum mottle.

In the treatment of imaging in terms of spatial frequency, quantum fluctuations are taken into account in the Wiener spectrum of the detector/recorder; if more quanta are utilized in forming an image of a given area this is equivalent to saying that the detector can be less efficient, for example, a slower finer-grain film, which will have a different Wiener spectrum.

8.8 INFORMATION THEORY

The technique of using spatial frequency as a parameter to study imaging processes has been shown to have further uses. In 1948 Shannon [19] developed a mathematical technique which showed that for certain types of communication system the information conveyed by a given signal can be calculated from considerations of the signal-to-noise ratio, and the term 'information theory' is now usually used for this technique.

The idea of information capacity as a measurable quantity may seem strange, but to take the photographic film case specifically, one can consider an area of photographic image as a series of separate cells each of area A, in each of which information is stored. If the film is coarse-grained the area of each cell must be larger, because of the need to prevent crosstalk between cells, so the number of cells per unit area must be smaller than on a fine-grain film, and the information capacity is less.

The simplest possible case would be if each cell was black or blank corresponding to a yes/no 'bit' of information, but in practice it is possible to distinguish between a series of different grey levels in each cell, so each cell can carry more than one bit of information. Due to noise, these grey levels must have a minimum separation, and with M grey levels and N cells per unit area, the information content is

$$C = N \log_2 M \qquad (8.18)$$

The unit of information is the binary digit, or bit, which corresponds to the amount of information given by a 'yes' or a 'no' answer, when both are equally probable. Fellgett [14], Linfoot [20] and Kanamori [21] extended the theory to photographic film and showed that the mean information content I recorded in an area H of emulsion is given by

$$I = H \iint \log_e \left[1 + \frac{S(u, v)}{N(u, v)} \right] du \, dv \quad \text{(bits)} \qquad (8.19)$$

where $S(u, v)$ and $N(u, v)$ are the signal and noise functions, respectively. Due to the isotropic properties of photographic film, measurements taken in one direction are sufficient to provide a complete description, so

$$I = 2\pi \int_0^f \log_2 \left\{ 1 + \frac{[R_0(f)]^2 P(f)}{W(f)} \right\} f \, df \quad \text{(bits per unit area)} \qquad (8.20)$$

where $P(f)$ is the input signal to the film, i.e. the intensity distribution arriving at the film surface.

There is one important limitation to equation 8.20 in this form. $P(f)$ is the input to the film, whereas the input to the radiographic system will consist of a finite number of X-ray quanta, with an inherent signal-to-noise ratio because of the random nature of X-ray quantum production, further modified by the transfer function of the system preceding the film. A comprehensive description of the information content of the whole system is therefore considerably more complex.

A useful method of using equation 8.20 is to consider the case where $P(f)$ tends to zero, as this represents a low-contrast image or a very small flaw in a specimen; this limiting value can be obtained because $(1/x) \log (1 + x)$ is a decreasing function of x when $x > 0$ and the upper limit is given by

$$4.36 \int \frac{[R_0(f)]^2}{W(f)} f \, df \quad \text{(bits)} \qquad (8.21)$$

Calculations of the information content of a high quality $5 \text{ in} \times 7 \text{ in}$ photographic print gave values of 5×10^7 bits and for a television single frame $(1/30 \text{ s})$, 30 000 bits.

Kanamori [21–23] applied information theory to calculate the information capacity of radiographic images. He claimed that a directly-exposed

Table 8.4 Information capacity of radiographic images using 70 kV
X-rays and Fuji KX film

Screens	Radiographic image (bits min^{-2})	Inspected by eye (bits min^{-2})
High-speed salt	159	159
Medium-speed salt	273	273
High-definition salt	837	635
No screens	58 000	951

radiograph contains 60 000 bits mm^{-2}, whereas with salt screens the
figure is a few hundred bits per square millimetre (Table 8.4). Also, the
useful information capacity of directly- exposed films is severely limited
by inspection by eye.

Kanamori used Wiener spectrum data which are not in good ac-
cord with other published values, and his information capacity
figure for a non-screen film is higher than would be expected from
Jones values [24] for photographic films exposed to visible light
($0.5 \times 10^4 - 3 \times 10^4$ bits mm^{-2}). Further, Table 8.4 suggests that to get the
maximum information out of a non-screen exposure, a fairly high view-
ing magnification is necessary. This is not in agreement with experience,
even for radiographs exposed to such low energy radiation as 70 kV
X-rays. It has generally been found that on fine-grain film some magnification ($\times 2 - \times 3$) is desirable, but on medium-grain direct-type films
comparable with Fuji KX, viewing magnification obscures rather than
increases the detail which can be discerned.

REFERENCES

[1] Halmshaw, R. (1979) *Brit. J. NDT.*, **21**(5), 245.
[2] Yokota, O. and Ishii, Y. (1978) *Jap. J. NDI*, **27**(1), 3.
[3] Morikawa, Y. and Koyama, Y. (1978) *Jap. J. NDI*, **27**(1), 35.
[4] Lamar, E. S. and Hecht, S. (1947) *J. Opt. Soc. Am.*, **37**, 531.
[5] Mizunuma, M. (1994) *Jap. J. NDI*, **43**(4), 215.
[6] CEN (1995) *EN-584-1*
[7] Jones, R. C. (1955) *J. Opt. Soc. Am.*, **45**, 799.
[8] de Belder, M. (1967) *Photo. Sci. & Engng.*, **11**(6), 371.
[9] Shaw, R. (1962) *Photo. Sci. & Engng.*, **6**(5), 281.
[10] Halmshaw, R. (1971) *J. Photo. Sci.*, **19**, 167.
[11] Stade, J. (1987) *J. Photo. Sci.*, **35**, 83; (1988) *J. Photo. Sci.*, **36**, 134.
[12] Halmshaw, R. (1973) CNAA, London *Ph.D. thesis.*
[13] Sturm, R. E. and Morgan, R. H. (1949) *Am. J. Roentgenol.*, **62**(5), 617.
[14] Fellgett, P. (1962) *J. Photo. Sci.*, **11**(1), 31.
[15] de Belder, M. (1969) *Proc. 5th Conf. Industrial Radiography*, Antwerp, Belgium. Gevaert Ltd.
[16] Saunders, A. E. (1976) *J. Photo. Sci.*, **24**, 96.
[17] Rose, A. (1948) *J. Opt. Soc. Am.*, **38**, 196.
[18] Rose, A. (1946) *J. Motion Pict. Engng.*, **47**(4), 273.

[19] Shannon, C. E. (1948) *Bell Syst. Tech. J.*, **27**, 623.
[20] Linfoot, E. H. (1956) *J. Opt. Soc. Am.*, **46**, 740.
[21] Kanamori, H. (1968) *Jap. J. Appl. Phys.*, **7**(4), 414.
[22] Kanamori, H. (1970) *Jap. J. Appl. Phys.*, **9**(2), 182.
[23] Kanamori, H. (1970) *Jap. J. Appl. Phys.*, **9**(11), 1378.
[24] Jones, R. C. (1961) *J. Opt. Soc. Am.*, **51**, 1159.

9

Interpretation of radiographs

9.1 INTRODUCTION

The basic purpose of radiographic inspection is to obtain information about abnormalities in the specimen. These abnormalities may be flaws in metal (e.g. cracks, inclusions), errors in assembly, wrongly-positioned components or dimensional errors, and may or may not be significant. An understanding of the image provided assumes therefore that an image of a 'correct' specimen is available for comparison.

In the case of flaw detection, in such applications as weld inspection, there is no need to actually provide a defect-free image, but for assemblies such a comparison image, taken at exactly the same orientation as the radiograph to be assessed, is essential. Interpretation may be said to have four stages.

1. Verification that the radiograph corresponds to the specimen or part of the specimen (identification markers) being examined.
2. Verification that the technique said to have been used has been used, and is appropriate to the specimen (codes of good practice, IQI readings).
3. Recognition of any artefacts. All artefacts which could be confused with genuine defects in the specimen should be identified and noted.
4. Identification of any flaws in the specimen, reporting the findings.

In the case of flaw detection in welds, castings etc., the fourth stage involves identifying the nature of the flaw (crack, inclusion, gas cavity etc.), estimating its dimensions and determining its location in the specimen. In practice many radiologists also have to report on the acceptability of the flaws which are discovered, in terms of an acceptance standard. Successful interpretation of industrial radiographs depends on intuition, as well as on information learned either by training or experience. First, it requires a sound basic knowledge of all aspects of the specimen being examined, such as the materials of which it consists, the positions of its component parts, and the types of defects likely to arise in it. Secondly, a knowledge of the basic principles of the radiographic technique used is essential, particularly the absorptive and scattering properties of the various parts of the specimen for the type of radiation used; interpretation cannot be left to non-radiographic personnel. Thirdly, it is necessary to have information in advance concerning the types of defects or other abnormalities likely to be encountered in the particular specimen, the manner in which their shape is likely to vary with the angle at which they occur with respect to the radiation beam, and the way their image varies in contrast and definition with their position in the specimen. An essential basis for sound interpretation is a wide knowledge of the characteristic appearance of defects, whether they are flaws in welds or castings, or distorted detail in assemblies, associated with the particular types of material or mechanisms in which they can occur. This knowledge can only be gained from experience in scrutinizing a wide range of specimens, radiographed and preferably correlated with sectioned and broken-down samples.

9.2 GENERAL ASPECTS

Viewing

A radiograph is examined by placing it on an illuminated viewing screen of appropriate luminance. It cannot be overemphasized that good film viewing conditions are absolutely essential to film interpretation. Holding a film up to the sun is grossly inadequate. Precise recommendations on film viewing conditions were given in detail in section 7.4.

Most film illuminators use one or more small Photoflood bulbs, which can be switched on for short periods by a separate switch. These bulbs have an opalized glass envelope which provides adequate light diffusion. For the examination of small areas of film of high density, a Photoflood bulb in a box with a restricted aperture on the front and no other diffuser is very useful. The radiograph may be a few centimetres from the glass of the bulb and the film is moved slowly across the aperture, which prevents any film scorching problems. Such a technique is satisfactory for one man examining a film, but it cannot be used for display purposes; used in a dark room, such a single Photoflood viewer is adequate for densities up to about four.

Attempts have been made to build similar restricted-aperture viewers for higher densities, up to five or six, using quartz-iodine or similar bulbs [1]. The major difficulty is that the bulb must not be brought up to full power until the film is in position, and the light must be instantly reduced or extinguished if the film slips out of position, otherwise there is danger to the eyes. There are also problems in protecting the film from heat. Such very high intensity illuminators are not therefore very practical. In addition to the high luminance requirements, the viewing light must be reasonably diffused, but need not be fully diffuse.

The radiograph must be adequately masked to eliminate glare emanating from around the film edges or from any local part of the film having particularly low density. Some of this masking can be done with adjustable blinds on the viewing screen, but a collection of strips of dark card for additional masking is almost essential.

The radiograph should be examined in a darkened room, with care being taken that as little as possible of the light reflected off the surface of the film reaches the film reader.

Density variations

It is a fundamental tenet of radiography that a thinner section of material absorbs less radiation and is therefore represented on the film by a higher film density, i.e. a darker area of image. All cavities in specimens and all inclusions of material of lower physical density than the parent material are therefore imaged as higher densities, that is, darker images on the film. An 'image' of less than background density must therefore be either an inclusion of higher physical density, or a surface feature (such as spatter on a weld) or a spurious occurrence.

The second basic tenet is that, except in projective magnification techniques, the images of defects are natural-size within about 10% magnification. The image of a cavity is a projection of a three-dimensional object on to a two-dimensional film plane, and some allowance must be made for this, but essentially the projection is based on a near-point source a considerable distance away. The only important exceptions to this are the images of very small or narrow flaws such as cracks, where the images may be spread out to several times the true flaw-width by unsharpness effects.

Artefacts

Radiographic film is subject to a number of artefacts, i.e. spurious images. Most of these are fairly easy to recognize and are unlikely to be confused with genuine specimen defects, but they occasionally cause confusion and misinterpertation.

The great majority of spurious markings on radiographs are due to faulty processing or careless handling; a very small number are due to imperfections in the original coating process by the manufacturers. Some of the more common artefacts are listed below.

1. Artefacts arising before exposure:
   (a) film mottle: due to the use of stale film;
   (b) radiation fogging: occurs when the film is stored too near a source of radiation, or when a film is inadvertently left in the exposure room during the exposure of another film;
   (c) light fog: caused by storage of film in a faulty storage box or bin; leaving the lid off the box; exposure to white light in a faulty darkroom or to the use of the wrong type of safelight or too strong a bulb in the safelight, or to the use of a faulty film holder; it is usually local but may be overall fog;
   (d) pressure markings: due to clumsy handling of the film when loading or unloading the cassette or film holder; they are often in the shape of dark or light crescents: if dark, they are caused by local pressure or bending the film after exposure; if light, local pressure before exposure;
   (e) scratch marks: usually caused by a fingernail or abrasive material;
   (f) static fog: has the appearance of branched and jagged fine lines; it is due to electric discharges on the surface of the emulsion when the film is removed rapidly from a tight wrapper (very rare);
   (g) high or low density fingerprints: caused when handling the film with greasy or chemically-stained fingers;
   (h) low density patches or smears: due to splashes of water or fixing solution on the film;
   (i) dark patches or smears: due to splashes of developer on the film;
   (j) radioactive spotting: occurs as very intense black spots, often with a halo around them; caused by radioactive contamination of the wrapping paper (now very rare);
   (k) light spots: dust particles between the film and intensifying screens.
2. Artefacts caused during exposure:
   (a) as 1 (b)–(d);
   (b) screen marks: due to contamination of the intensifying screens with chemicals, or to defects in the screens such as cracking or buckling, or to the presence of dust.
3. Artefacts caused during processing:
   (a) as 1 (c), (d), (g) and (j);
   (b) air bells: these are shown as discs of lower density, due to air trapped on the surface of the emulsion, usually during the early stages of development, due to insufficient initial agitation;
   (c) patchiness or streaks: due to inefficient agitation during development, or failure to agitate in the rinsing bath (quite common);
   (d) oxidation (aerial) fog: caused by excessive exposure of the film to air during development;
   (e) neighbourhood effects: these are shown as streaks from part of the image and are due to insufficient agitation during develop-

ment, especially when the image contains a sharp boundary separating areas of high and low density;

(f) reticulation: this has the appearance of leather grain and is due to rupture of the emulsion caused by great differences in temperature between successive processing solutions (now very rare);

(g) drying marks: due to drops or streams of water remaining on the surface of the film after it has been partially dried, they often occur when attempting to dry films rapidly at a high temperature in a drying cabinet (quite common).

Some of these artefacts are exceedingly rare and represent gross mishandling of films. By far the commonest are:

1. pressure marks, which are usually easily recognized and not likely to be mistaken for defects;
2. patchiness or streaks, which can be exceedingly difficult to eliminate completely;
3. drying marks, which are effectively another form of patchiness;
4. neighbourhood effects, which are again not likely to affect interpretation;
5. dust particles on the film or screens.

It is frequently possible to identify spurious images such as scratches and drying marks by looking at the two emulsions of the film by reflected light, i.e. by viewing each film surface in turn at near glancing angle. If the image in question can be seen on one emulsion but not on the other, it is almost certainly spurious. In difficult and critical cases it may be necessary to take a duplicate radiograph.

9.3 WELD RADIOGRAPHS

The most widely used welding process to which radiographic inspection is applied is the metallic-arc fusion butt weld. Some other special welding processes have particular characteristic defects, but most welding defects are common to several welding processes. The International Institute of Welding produced a 'Classification of defects in metallic fusion welds' [2] and the recommended terms used in this document will be used here.

Considering fusion butt welds, the possible defects can be classified below as surface or internal defects.

1. Surface defects (Fig. 9.1)
   (a) undercutting
   (b) incompletely-filled groove
   (c) dressing, chipping, grinding marks
   (d) spatter

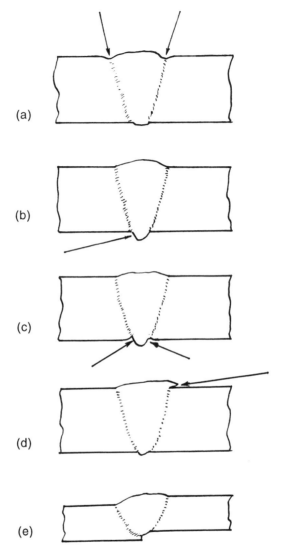

Fig. 9.1 Various surface defects in butt welds. (a) = undercutting; (b) = excessive underbead; (c) = shrinkage grooves; (d) = overlap; (e) = misalignment.

- (e) excessive underbead (excessive penetration)
- (f) shrinkage groove
- (g) overlap: an overflow of weld metal on to the surface of the parent metal
- (h) misalignment
- (i) crater pipe: due to shrinkage at the end of a weld run.

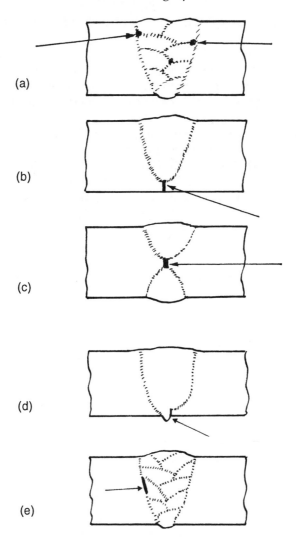

(a)

(b)

(c)

(d)

(e)

Fig. 9.2 Various welding defects. (a) = slag inclusion; (b) = lack of root penetration; (c) = lack of root penetration in a double-vee weld; (d) = lack of root fusion; (e) = lack of side-wall fusion.

2. Internal defects (Fig. 9.2)
   (a) Blowhole: a large cavity due to entrapped gas. This term is conventionally applied to cavities exceeding 1 mm diameter, and such cavities are easily detected by radiography.
   (b) Gas pore: a small cavity due to entrapped gas. Similarly, this term is conventionally applied to cavities less than 1 mm diameter.

(c) Porosity: a group of gas pores, often lying in strings, i.e. linear porosity.

(d) Pipe or worm hole: this is an elongated cavity, and is frequently end-on to the X-ray beam, so that the image is circular.

(e) Inclusion: slag or other foreign matter trapped during welding. These occur as isolated inclusions, clusters of inclusions, or lines of inclusions.

(f) Incomplete root penetration: a gap, sometimes intermittent, easily detected and identified by radiography because of its machine-edge straight-line faces.

(g) Lack of fusion: lack of adhesion between weld and parent metal or between runs of weld metal. Usually lack of side fusion, i.e. weld-to-parent metal or lack of root fusion is referred to, but lack of inter-run fusion is also possible. The actual gap between the unfused faces may be very narrow. If there is lack of root penetration at an unfused root face there is obviously also a lack of root fusion, but it is possible to have one root face fused to the weld metal and the other face unfused.

(h) Lack of sidewall fusion: this is not easy to detect by radiography, unless the radiation beam is directed along the sidewall or there is associated slag.

(i) Cracks: these may occur by tearing of the metal when it is in the plastic condition (hot crack) or by fracture when cold (stress crack). In some materials cracks can occur as a delayed process long after welding is complete, usually in the heat-affected zone. Cracks are often further differentiated as transverse, longitudinal, heat-affected zone, crater etc. As a defect, cracks are a special category in that they can propagate under stress, so that a small crack may be inherently as serious a defect as a large one from the point-of-view of strength; lack of root fusion and lack of sidewall fusion are both crack-like defects to which the same comments apply.

(j) Undercutting: on a radiograph, this shows as one or two moderately straight, broad lines, parallel to the line of the weld seam. These lines may be continuous or intermittent, are variable in width, and have very blurred edges; this last feature usually distinguishes them from internal slag lines, which typically produce sharper images. If there is any doubt, a visual inspection will confirm the diagnosis.

(k) Shrinkage grooves: these are surface grooves on the sides of the underbead, similar to undercutting but with a narrower spacing.

Misalignment: if the plates are misaligned a very characteristic radiographic image is produced which consists of a straight-edge step, not a line.

Apart from overlap, surface defects are all easily detected by radiography but should also be detected visually if both weld surfaces are accessible. A practical problem of considerable importance in radiography is that the finished weld is often left with a cross-section similar to Fig. 9.1(b), with extra metal on both sides, and this can cause an excessive density difference between the centre of the weld and the parent plate, making interpretation more difficult.

Gas pores, when small, are spherical and so are seen on radiographs as dark spots with diffuse edges; they may occur singly, or in lines, or clusters, or be generally distributed. They cannot be mistaken for any other weld defect. If the porosity is widely distributed, it is important to remember that the film will show pores at different distances through the metal superimposed in the image and will give an impression that the pores are more frequent and closer together than they really are. An important problem with heavy porosity is the possibility that it might obscure the image of a much more serious defect such as a small crack.

It is becoming common to describe porosity in percentage terms. For example, 1% porosity means that the total volume of all the gas pores in the weld metal is 1% of the volume of the weld metal.

Isolated gas cavities can occur in a wide range of sizes: there can be small-diameter cavities extending a very large distance through the thickness (Fig. 9.3), or cavities elongated along the length of the weld bead. They are not, therefore, always spherical. It is possible for the image of a wormhole to look like a gas pore except for the extra density in the image.

Inclusions of solid material occur as isolated defects, as clusters or in lines, and may consist of flux, slag, oxide or as particles of foreign metal. The first group are recognized on radiographs as irregularly-shaped dark images. The fact that the image is darker than the background indicates that the defect consists of material of lower physical density than the parent metal, and the irregular shape indicates that the cavity is unlikely to be gas-filled. The commonest form of inclusion is the line, or lines, of slag often called 'tramlines'. In welds made with a tungsten or copper electrode, particles of the electrode can be trapped in the weld metal. These are particles of higher density than the weld metal and so are seen as low-density images on the film.

Some modern electrodes contain a barium salt in the coating and if this material forms a constituent of a slag inclusion, it may be sufficiently absorbent of the X-rays to produce images of the inclusions which are photographically less dense than the surrounding metal. This will probably occur only on thin welds radiographed with low energy X-rays, but in theory an inclusion containing some barium salt could have an X-ray absorption exactly matching steel, for a particular X-ray energy, and so become undetectable [3].

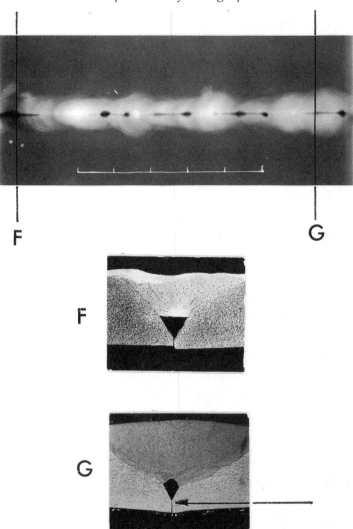

Fig. 9.3 Radiograph and cross-sections of weld in 6 mm steel. The cross-sections are magnified × 4 compared with the weld and show lack of root penetration and associated gas cavities.

Lack of root penetration is associated with the root faces and is unmistakable on the radiograph because of the very straight machined edges, often well-separated. If the root faces are very close together and not fused in the welding process, this is usually called 'lack of root fusion', but 'lack of root penetration' would be equally correct. Figure 9.3 shows examples of this defect, as well as the large isolated gas cavities which are sometimes associated with it in thin welds.

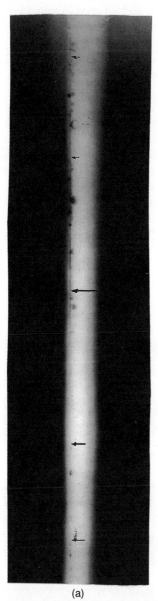

(a)

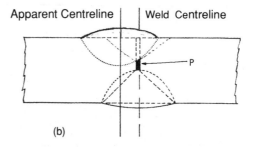

(b)

Fig. 9.4 Weld in 12 mm steel pipe, welded in two passes by submerged arc welding; Double-vee preparation. (a) = radiograph showing lack of root penetration and associated gas cavities; (b) = cross-section of weld.

Misalignment of the two sides of a weld (Fig. 9.4) can lead to some strange-looking defects such as lack of root penetration which is apparently off-centre. The arrowed defect is so straight, and has some associated gas cavities, that it is very typical of lack of root penetration, but in the wrong place. The cross-section shows the explanation. The weld bead on one side was well off the true weld centreline and had only partly removed the lack of penetration from the other half of the weld. The film had been placed on the top side of the weld (as drawn) and this weld bead showed prominently.

Cracks show on radiographs as thin dark lines with a characteristic wavy or broken appearance (Fig. 9.5). This appearance is caused by the twisting of the crack along its length, which presents a varying width to the X-ray beam. Cracks in welds can be anything from tens of centimetres long (longitudinal stress cracks, Fig. 9.5(a) and (b)) to 1–2 mm long (crater cracks, Fig. 9.5(c)). The ability to detect a crack depends very much on the crack opening width and its angle to the radiation beam: in thin welds, less than 10 mm thick, where the IQI sensitivity is 1–2%, crack sensitivity is high and very small cracks can be detected. On thicker specimens, over 80 mm, when high energy radiation has to be used, tight cracks are less likely to be found because of the greater total unsharpness, which spreads out the image. Nevertheless, many cracks are detected.

It is important to remember that if a crack is oriented at an angle to the X-ray beam, its image will not be narrow and contrasty, but broad and

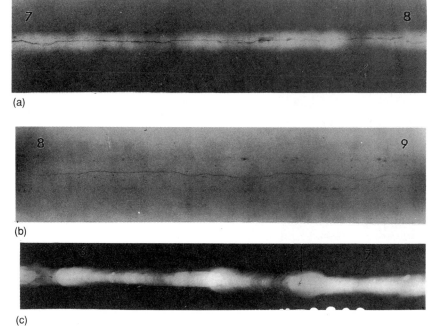

(a)

(b)

(c)

Fig. 9.5 Radiographs of cracks in welds. (a), (b) = longitudinal cracks; (c) = crater crack.

faint. It is very unusual for a crack in a butt weld to be at a considerable angle to the X-ray beam, but this can occur with other weld constructions. A case which causes difficulties in interpretation is the basal crack (i.e. the crack originating at the root of a weld) when the root bead has not been ground off. The edge of the root bead gives a sharp 'edge' image on the radiograph, and the crack image is in this 'edge' image; one has to decide whether the image is a crack or merely a step on the surface. The only way, if there is uncertainty, is to grind a small length of the inner weld surface smooth; if there is a crack, the image of the crack will persist (this grinding does not affect the serviceability of the weld, if there is no crack).

The interpretation of an image of partial penetration welds is especially difficult, and if another NDT method such as ultrasonic testing is possible, this is preferable. Cracks may be found by radiography in such welds, but there is a tendency for the crack image to be obscured by the image of the non-penetration gap. Similarly, radiography of fillet welds is not usually satisfactory and other methods of NDT are preferable, when practical. Radiography is sometimes used to examine electron-beam welding, where the commonest serious defect is an unfused or partly-fused weld, caused by the electron beam wandering off the line of the joint during welding. This is shown on the radiograph as a very fine straight line of 'lack of root fusion'; it can easily be missed if the alignment of the joint and X-ray beam is not accurate.

Most of the defects so far described can occur in any welding process, in any metal, but in spot-welding certain special defects occur. During the formation of a spot-weld in light-alloy sheet, where copper or zinc is used in the alloy, the alloying constituents tend to migrate into zones which are shown on the radiographs as rings of different density. It has been claimed that the image of these rings can be used as a criterion of weld quality. A thin, well-defined dark ring outlines the diameter of the fused nugget, and outside this a light corona corresponds to the heat-affected and welding pressure zone; these two rings should be clearly visible. A smaller diameter of dark ring corresponds to a reduction of penetration; greater radial thickness of the light zone corresponds to reduction of penetration, as does less differentiation of the two zones. In the case of imperfect bonding, the ring may be non-existent. For this type of work a very good radiographic technique, using fine-grain film, is essential. The interpretation of indications detailed above do not apply to spot-welds in steel plate.

Diffraction mottle

On radiographs of thin stainless-steel welds, a narrow diffuse line is sometimes shown on the centreline of the weld, which is too blurred in

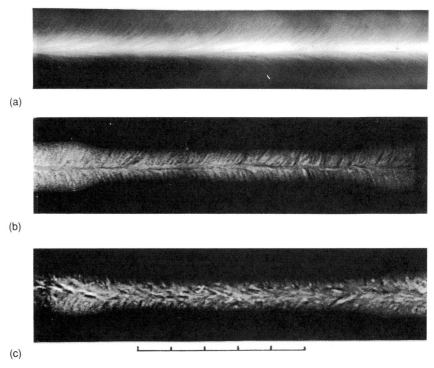

(a)

(b)

(c)

Fig. 9.6 Radiographs showing examples of diffraction mottling in thin austenitic steel welds. Each division in scale represents 1 cm.

appearance to be 'lack of root penetration'. Sometimes a 'herring-bone' pattern is shown (Fig. 9.6). The latter is also common on weld radiographs of light alloy welding.

These images have only occasionally been investigated and it is still not fully established whether the cause is partly due to variations in metal structure across the weld, causing differences in radiation absorption (for example, chromium depletion in individual grains in austenitic steels), or diffraction effects due to oriented grains in the metal. Watanabe [4] investigated the formation of herring-bone patterns on the radiographs of thin austenitic (18:8) steel welds and clearly showed that in his specimens the pattern was caused by diffraction. The pattern changed markedly with a change in specimen-to-film distance, or with a change in X-ray beam angle, or with a change in X-ray kilovoltage. He also concluded that the amount of diffraction mottling is controlled by the δ-ferrite content: if the δ-ferrite content is low the diffraction pattern is more pronounced. In a given specimen the diffraction pattern disappears when the specimen is heat treated, which suggests that the pattern is a characteristic of a cast structure. Watanabe also showed that as the contrast of the diffraction pattern increases there is some loss in mechanical strength. The main problem with diffraction patterns in interpretation of weld radiographs is when they sometimes have a similar appearance to a weld defect, but this is not often the case. The centre line on the weld on Fig. 9.6(b) might be mistaken for a line of incomplete penetration, but the image is not sufficiently sharp. Also, in diffraction images, a dark line is usually accompanied by an adjacent white line. Finally, the accompanying herring-bone pattern is highly characteristic of diffraction (see next section).

9.4 RADIOGRAPHS OF CASTINGS

There are some defects which are common to most casting processes, for both metals and non-metals, and some special defects which are characteristic of particular metals or processes. Castings are made in an enormous range of sizes, from a few grams in weight to nearly 100 tonnes, and they vary in shape from simple to extremely complex forms. The names given to various casting defects vary considerably between different countries, and even between different areas of the same country. The terminology used here for internal defects is taken from British Standard BS 2737: 1956.

The first group of defects are voids produced by entrapping of gas evolved from the metal, by entrapped air, or by shrinkage of the metal. The individual types of this defect are described below.

1. Microporosity. This is a very fine form of the defect in which very small cavities occur, usually around the grain boundaries. If the cavities link up into a fine, three-dimensional network some authorities call it microshrinkage. However, if the cavities are discrete the term microporosity is used. This usage retains the association between 'porosity' and 'pores', but BS 2737 and American nomenclature prefer to use 'porosity' for something which is porous, i.e. has a connected passage through the thickness. This confusion has never been eliminated in English. On the radiograph, on a thin section, a fine network pattern may be discerned, but on a thicker section the individual cavities are superimposed and the radiographic image is a general mottle or cloudy effect. This defect is common in aluminium alloys but comparatively rare on radiographs of cast steel. In magnesium-base alloys it sometimes occurs in layers, shown as dark streaks on the radiograph.
2. Porosity. If the individual pores are visible on the radiograph as small dark spots less than about 1 mm diameter the defect is called porosity or sometimes pinhole porosity, and if the radiograph shows larger dark circular images of smooth outline these are usually called gas holes. A particular form of gas hole, elongated to

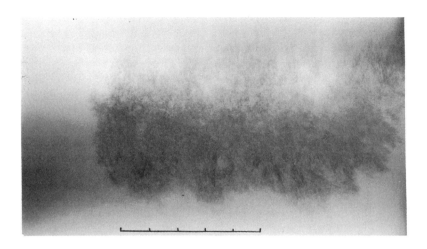

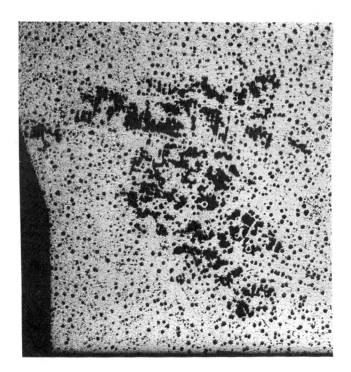

Fig. 9.7 Radiograph and cross-section showing sponginess in cast-iron. Each division on the scale represents 1 cm.

extend to the surface, is called a wormhole. These are easily recognized on a radiograph because they occur in clusters, and there are always some in the cluster which are angled with respect to the radiation beam and which can be seen to be tube-like.

3. Sponginess. If there is a system of interconnected small cavities of a slightly coarser form than microshrinkage, this is called sponginess. In thin sections the pattern of the cavities can be discerned on the radiograph, but on thicker sections it may be shown only as a patchiness and be difficult to identify. It is caused by metal shrinkage, is very common in cast-iron (Fig. 9.7), where it has a characteristic appearance, in stainless-steel and in copper alloy castings. Except on very thick sections, it is easily identified. The distinction between microshrinkage, sponginess, and filamentary shrinkage is largely one of size; they are essentially the same nature of defect.

4. Airlock. Sometimes air is trapped in the casting mould by the inflowing molten metal, and these cavities are recognizable on the radiograph as dark areas with a smooth outline, but are not generally circular like a gas hole.

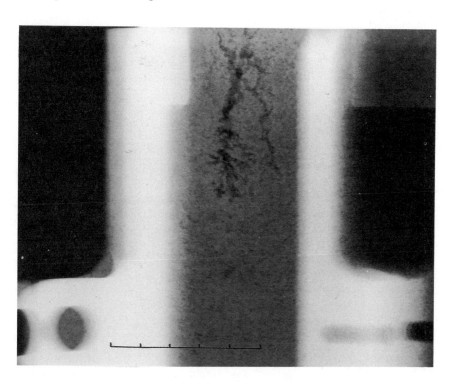

Fig. 9.8 Radiograph showing shrinkage in an aluminium-alloy casting. Each division on the scale represents 1 cm.

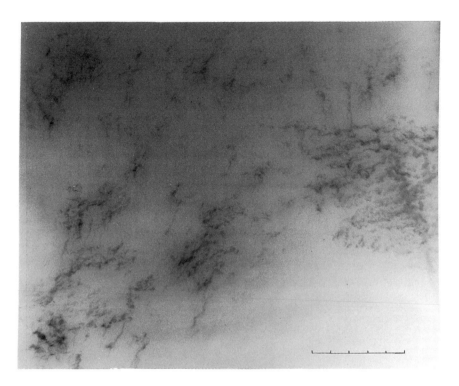

Fig. 9.9 Radiographs showing centre-line shrinkage in cast steel. Each division on the scale represents 1 cm.

5. Shrinkage. These are cavities caused by contraction of metal and show on radiographs as piping, filamentary, or a fine branching network (Fig. 9.8). They are usually extensive cavities and their form depends on the metal thickness and the rate of cooling. They are most typical in medium and large castings. In filamentary shrinkage there is usually a tree-like structure with trunk and major and minor branches, and the finer arms may degenerate into sponginess. A special form of this defect which occurs in large areas of uniform thickness such as cast plates, often occurs on a plane near the centreline of the wall thickness and is called 'centreline shrinkage' (Fig. 9.9).
6. Cracks. These occur as hot tears or stress cracks. Hot tears have a characteristic radiographic appearance (Fig. 9.10). The radiographic image is a dark ragged line, often discontinuous, and this nature distinguishes the defect from piping or filamentary shrinkage. Stress cracks show on radiographs as dark wavy lines, which are smooth and change gradually as the cracks twist slowly.

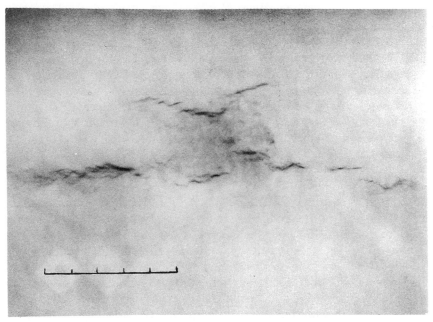

Fig. 9.10 Radiograph showing hot tear in a steel casting. Each division on the scale represents 1 cm.

In castings there is often little chance of predicting the orientation of a crack, and as a consequence, radiographs may show cracks where the radiation beam is at a considerable angle. In this case the image of the crack rapidly becomes of much lower contrast and is spread out into a rather faint broad line; it retains its characteristic wavy appearance and can usually be recognized because of this.

7. Cold shut. This is a defect formed when two streams of molten metal fail to unite. On the radiograph it is usually seen as a dark line, and may be difficult to distinguish from a crack. Unfused chills in castings produce similar images but can usually be recognized by the characteristic shape of the chill.

8. Unfused chaplets. Chaplets are devices for maintaining the position of mould cores and usually consist of small bars with end-plates; the bar should fuse into the casting metal, but sometimes does not fuse properly and sometimes produces gas and inclusions.

9. Segregations. This is a condition resulting from local concentrations of any of the constituents of an alloy. It may be general, localized or banded. It is also possible to have local segregation in which shrinkage or hot tears are filled with segregate. General segregation can be

a normal condition of certain alloys and may show as an overall fine pattern or mottle. On thick sections segregation may be difficult to distinguish from microporosity, but generally the pattern is slightly coarser.

10. Inclusions. The commonest inclusions in steel castings are sand, slag or dross, trapped in the molten metal. These are of lower physical density than the parent metal and so are seen on the radiograph as irregular dark images which can be distinguished from gas holes by their shape. Slag usually gives a more rounded image than sand or dross, which is trapped as a solid material.

Diffraction effects on casting radiographs

In light-alloy castings the diffraction pattern, when it occurs, is more often seen as a mottle or a pattern of irregular spots, and this can be less easy to distinguish from small porosity or microshrinkage. If it is possible to take a second radiograph the problem can usually be resolved, as a small tube-shift or change in radiation beam angle causes a large change in the diffraction pattern, and also the diffraction spots can often be seen to extend outside the image of the specimen. An alternative technique is to take the second radiograph with the film deliberately spaced away from the specimen by 1–2 cm. Glaisher *et al.* [5] obtained diffraction mottle in specimens of coarse-grained pure aluminium and in aluminium/4% copper alloy castings, and showed that the dark spots were Laue spots and that there were lower density local regions in the transmitted beam where some energy had been diffracted out. They also suggested that with an alloy containing more than one phase, radiographs of very thin specimens could show up the phase distribution due to differential absorption; this would not be a diffraction effect.

Mottle was obtained on the radiographs of 9% silicon/4% copper/aluminium alloy castings by van Horn [6], but he claimed that the copper content and the microsegregation of the copper-rich phase also affects the mottle, and there has always been some unresolved doubts as to whether slight differences in radiation absorption between individual large grains in the metal might be a secondary cause of mottle.

Occasionally diffraction effects can produce some bizarre images [7,8]. The Author met one case where star images each with arms 35 mm long appeared on gamma-radiographs taken with Ir-192 of a 70 mm thick austenitic steel cast ring, which proved to have a very large grain structure; the stars were about 60 mm apart and occurred on every radiograph taken with the source at the centre of the ring and the film outside, on one particular specimen (Fig. 9.11).

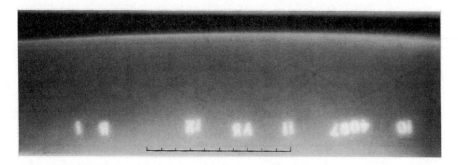

Fig. 9.11 Gamma-radiograph of 70 mm thick cast austenitic steel, showing diffraction 'images'.

Generally, however, diffraction images are seen only with thin specimens and low energy X-rays.

One of the important problems in interpretation of casting radiographs is the determination of the location of a particular defect through the thickness. Thus, shrinkage piping at the centre of a cross-section is likely to be comparatively harmless, but if there are filaments reaching to either surface, it becomes a much more dangerous flaw. Stereoradiography, as described in Chapter 6, can be used to locate flaws accurately, but is comparatively slow and often sufficient information can be obtained either from a knowledge of the casting process or from radiographs taken at different angles to cover different sections. Often, on castings, two radiographs with beam directions at 90° are taken, and these, examined together, give complete information on both defect size and location.

Special problems arise in the radiography of small ferrous castings in which section-changes tend to be more rapid, but the main problem is in producing good radiographs, rather than in interpretation.

Most of the defects occurring in light-alloy castings are easy to recognize but there are some special problems in interpretation. Microporosity, microshrinkage, sponginess and segregation are much more common and not always easy to identify, or to separate from diffraction mottling. In certain alloys it has been claimed that segregation, as seen on the radiograph, is an indication of a satisfactory metallurgical structure, so there is obvious need not to confuse it with sponginess. In light-alloy castings a surface coat of paint containing some dense material can cause considerable confusion in interpretation; the paint can fill a surface crack or cavity with material which is more absorbent to radiation than the parent metal, and so produce images of reversed tones. Similarly, a sand inclusion in a magnesium-alloy casting may be shown as a lighter area on the radiograph.

Radiography is nowadays being applied more extensively to copper-based alloys. The defects which occur are similar to those already described in steel castings.

9.5 NON-METALLIC MATERIALS

The application of radiography to a wide range of non-metallic materials has expanded rapidly, as such materials are used for critical structures and purposes. One large field of application is to explosives, which are produced in both cast and extruded forms. Rocket propellant is frequently formed by an extrusion process and is subject to a number of special defects in addition to the usual ones. As high explosive is usually cast inside metal containers, the usual casting defects are much more common, because it is not always possible to use the necessary headers and feeders to produce a good casting. Interpretation, again, is not difficult, and typical defects are large and easily visible on the radiographs.

9.6 REFERENCE RADIOGRAPHS

In the USA there have been extensive attempts to codify the interpretation of casting radiographs by the use of collections of reference radiographs. These have been marketed by ASTM and there are series for X-rays, gamma-rays, megavoltage X-rays, for steel castings and for castings in other metals, for thicknesses up to 12 in. As an example, in the medium thickness collection, internal casting defects are divided into seven groups: sand and other inclusions, gas and blowholes, shrinkage, chaplets, chills, cracks and hot tears, and there are radiographs illustrating different degrees of severity, sometimes five degrees, of each. These are original or copy films, usually taken from specially-produced test specimens. The intention is that these reference radiographs shall be used as acceptance standards, and there is guidance given on classes of castings according to the severity of service conditions; although this usage is controversial, the value of these sets of radiographs for interpretation purposes is obvious. No radiographic techniques are included with the films, but there is an assumption of a good technique.

The appearance of a defect on a radiograph depends on several factors, such as the orientation of a defect relative to the radiation beam, the radiographic technique, the total specimen thickness, the position of a defect through and the thickness, but no set of reference radiographs has attempted to cover all these variations for any particular defect.

In 1952 the International Institute of Welding produced a set of reference radiographs of welds in steel. These illustrated the various possible weld defects and on most defects there were illustrations of different degrees of severity, which were identified by colours:

- Black, a weld containing only a few small scattered gas pores;
- Blue, very small amounts of gas holes, slag inclusions or undercutting;
- Green, small amounts of gas holes, slag, undercutting or incomplete penetration;
- Brown, moderate amounts of these defects, or of lack of fusion;
- Red, gross amounts of these defects, or containing cracks.

Thus there were, for example, illustrations of four degrees of severity of some defects such as slag inclusions.

In 1985 a new collection illustrating weld defects and cross-sections through the welds was issued. Similar collections of reference radiographs of welds have been produced in the USA and Germany.

The IIW collection was intended only for educational purposes, but some organizations have also used it for acceptance standards. This is a misuse of these reference radiographs, as there was never any intention that films of the same colour for two different weld defect-types should be equated in severity.

It is very convenient to illustrate different degrees of severity of distributed defects such as porosity, microshrinkage, microporosity etc., by reference radiographs rather than in words, but from the viewpoint of acceptance of defects, each type of defect needs to be treated separately.

9.7 ACCEPTANCE STANDARDS FOR DEFECTS

There has been a slow build-up of quantitative data from the results of mechanical tests on defect-containing specimens, for example, fatigue testing of welds, and theoretical predictions of 'safe-life' have been developed from fracture toughness considerations. It is important to remember that acceptance levels may be set from two quite different standpoints:

1. knowledge that a particular defect might cause failure, or propagate to a dangerous size;
2. knowledge that a particular defect or group of defects should not be present if the workmanship was up to normal standards, leading to the suspicion that a particular specimen is sub-standard and therefore suspect.

In the absence of any specific acceptance standards from the purchaser, the following guidelines for castings and welds can be used. For castings:

- cracks and hot tears are unacceptable;
- sponginess is unacceptable;
- cold shut is unacceptable;
- unfused chill or chaplet is unacceptable;

- filamentary shrinkage, if recognizable as centreline shrinkage, is acceptable; general filamentary shrinkage and piping are the most difficult defects about which to generalize, each individual case must be treated separately;
- sand inclusions are not important from the point-of-view of strength, but can affect machining; and are often machined out during finish-machining.

It should be emphasized that these are only guidelines, for use in cases where the customer has not produced any acceptance standards.

For welds, CEN Standard EN–25817:1992 (ISO–5817:1992), 'Arc-welded joints in steel – guidance on quality levels for imperfections', is now used throughout the EC. This lists three quality levels: D, moderate; C, intermediate; and B, stringent. The acceptability of weld flaws is specified under these headings. No cracks except crater cracks are permitted, and crater cracks are only permitted in class D. Lack of fusion and lack of penetration are not permitted except when shallow and in very short lengths. Porosity is permitted up to 4% in class D and 1% in class B with a provision on the minimum size of individual pores. There are similar detailed sizing limits for all other welding defects. It is stated that these provisions apply to welds 3–63 mm thick and 'may well be applicable outside this range'.

Two problems are apparent in applying this standard for radiographic applications:

1. in some cases a through-thickness height of the flaw is quoted, which is not easy to determine by radiography;
2. some weld configurations, such as some fillet welds, are not easy to examine satisfactorily by radiography.

Acceptance standards, in general, should be set for individual items, for known environmental conditions. Some British standards, for example BS:3351–1971 and BS:4515–1969, contain acceptance standards.

9.8 ASSEMBLIES

So far as interpretation is concerned, there are two basic rules for the radiography of assemblies.

1. Always choose the orientation of the radiation beam which gives a simple image. If there are plane faces these are best shown as straight lines by having the beam directed along these faces. Oblique shots should be avoided whenever possible. To this end, a large source-to-film distance to reduce distortion and produce an image which corresponds most nearly to a cross-sectional drawing is advantageous.

2. A 'master-radiograph' of an assembly which is known to be correctly
   assembled is always desirable and greatly simplifies interpretation.

Most mistakes in interpretation of assemblies arise from attempts to
interpret a radiograph of an unknown object, when one is trying to
reconstruct a three-dimensional object from a two-dimensional image.

If it is necessary to examine unknown objects, one should, as a minimum, have two radiographs taken in mutually perpendicular directions,
unless the specimen geometry makes this quite impossible. On most
radiographs of assemblies there is a considerable film-density range and
it should be remembered that on parts of the image of density less than
1.0, there is considerable loss of detail sensitivity. A small-area high-
intensity viewer is of considerable value for examining local areas of
high film density. Where it is necessary to show the outer edges of an
object accurately, particularly cylindrical objects, there is very considerable scope for special techniques using filters, higher energy radiation,
single-emulsion films and curtailed development. Some of these methods can produce an enormous improvement in radiographic quality and
so simplify interpretation.

9.9 REPORTING RESULTS

A radiographic report should contain all the relevant information given
by the radiograph, together with sufficient of the nature and history of
the specimens to permit unequivocal identification at any later date. The
use of a standardized report form enables all the relevant data to be
recorded efficiently. Most organizations also record the technique details
in sufficient depth to allow radiographs to be taken of subsequent specimens of the same pattern.

An important aspect of any report is the method of describing the
location of radiographs over a large specimen, or the orientation of the
specimen when several radiographs are taken on the same specimen.
Most radiographers use some sort of 'shorthand' code of lead letters on
their films, to identify angles and positions, and often this code is not
adequately recorded.

Finally, the crux of the radiographic report is a precise description of
all the significant defects found in the specimen, in respect of nature,
extent, location and, if the reporter is competent, with an opinion of the
significance of such defects.

Any such assessment pre-supposes a knowledge of the service conditions and function of the specimen, as well as the effects of various
defects on strength. It may be argued that the radiologist's function ends
at a description of the defects, while it is the function of the design
engineer to pronounce on the acceptability of defects. Unfortunately the

designer is not necessarily available for consultation when required, and further, may not be competent to interpret radiographs, so whatever may be the ideal state of affairs, it is often left to the radiologist to pronounce on acceptance or rejection.

9.10 IMAGE DIGITIZATION FROM FILMS

Many insurance companies require radiographs to be stored for 10–20 years and this is not easy without severe deterioration unless the film has been processed very carefully. In addition, good storage conditions and adequate storage space are necessary.

It is possible to record the image on a radiograph by scanning the film with a very small light-spot, converting the reading to digital form and storing this on tape or disc. The first major use of image digitization from film has been for image storage, but there are also great potential uses for image enhancement and automated image interpretation (Chapter 12).

For storage purposes it is necessary to relate the size of the scanning light-spot to the resolution of the image on the radiograph (usually expressed in terms of the total acting unsharpness). Also, the grey-scale of the data storage system must be related to the density range of the radiographic image: 8 bits corresponds to 256 grey levels and 12 bits to 4096 levels. To cover a film density of 0–4 in 0.01 density steps would require 400 grey levels, but a typical weld radiograph is unlikely to have a density range of more than two (3.5–1.5), which is only 200 grey levels. More levels may be needed for image processing.

The transmitted light intensity through the film, coupled with the spot area, must be such as not to introduce any signal-to-noise limitations. A laser light beam or a CCD camera is usually used for scanning [9,10].

Low voltage X-radiographs with a very small total unsharpness can be shown to require a scanning spot within the range 30–50 μm in diameter if there is to be no loss of information, while a 100 μm spot should be adequate for most other radiographs. Density variations as small as 0.01 should be separable.

A pixel size of 50 μm corresponds to 6000 pixels across a 300 mm film image. Halving the resolution reduces the data volume by a factor of four.

The data are usually stored on an optical disc (write once, read many times (WORM)) either as uncompressed or slightly compressed data. A small amount of data compression need not lose information. The capacity of a 12 in optical disc is 6.4 Gbyte, which corresponds to several thousand radiographs, and not all parts of a film image need to be stored at full resolution.

It is essential to be able to re-examine the image on a monitor screen with the same image quality as the original film, and this is at present usually done with either a 1000 or 2000 raster-line image [9,10].

Some recommendations for the requirements for image digitization are detailed in US Nuclear Regulatory Commission document NUREG–1452(1992) and for standardization in ASTM E–1475–1994.

9.11 PROBABILITY AND RELIABILITY

It is well known that human error is possible in any image interpretation process, whether this is due to misinterpretation, lack of skill or tiredness etc. In the limit when one is looking for images near the limit of discernibility, probability becomes an increasingly important factor: a sharp clear image is seen with a near 100% probability, a fainter, smaller image with less certainty. Several investigators have, in recent years, taken a large collection of flawed specimens, examined these with a range of different NDT techniques, and produced probability of detection (POD) curves which purport to compare the different techniques used. Usually the investigation has compared radiography and ultrasonic testing [11–14].

The practical problem is that even if the experiment is limited to weld flaws, there are still six major types of flaw, as well as different sizes in each flaw group, so that to get a statistically significant number of each flaw type in each group, a very large number of specimens is necessary. In addition, in radiography different flaw sensitivities are possible with different radiographic techniques, and usually only one radiographic technique has been used.

Figure 9.12 shows the type of curve which has been produced for different weld flaws for one radiographic technique, based on a total of 700 flaws and a total of more than 3000 interpretations by six inspectors. The POD curves are plotted in terms of flaw height, but there is no information on how this height was determined. The curves show, qualitatively, what one would expect. As the flaw gets larger the probability of detection increases; cracks are detected with less certainty than volumetric flaws such as porosity and slag inclusions.

In radiography, as the flaw gets smaller and the image is nearer the limit of discernibility, it is still possible to detect an image but to be uncertain of the nature of the flaw, due to image unsharpness etc. Thus, a true crack might be interpreted as a slag inclusion, and in terms of a POD curve for cracks, such as curve E in Fig. 9.12, be classified as 'not detected'.

A different method of assessing reliability has originated from signal detection theory for the detection of a weak signal in a noisy background. This is the production of receiver operating characteristic (ROC) curves. So far as is known, these have not yet been applied to industrial radiography, but attempts have been made to apply them to diagnostic medical radiography [15].

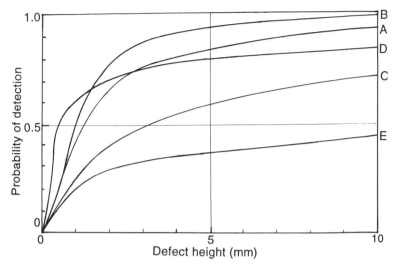

Fig. 9.12 Probability of detection curves obtained for one radiographic technique for five different types of weld flaw [12]. Curve A = porosity; Curve B = slag inclusions; Curve C = lack of fusion; Curve D = incomplete penetration; Curve E = cracks.

9.12 COPYING RADIOGRAPHS

Occasionally there is need to produce duplicate radiographs or copies, for storage, reports, second opinions etc. Several methods are possible, depending on the density range of the original film. If the original has a long density range, undoubtedly the best method is to take a second radiograph of the specimen whenever this is possible. Alternative methods are given below.

1. To make a normal print on photographic printing paper from the original. This is the most economic method, but the print will have reversed tones from the original (i.e. a normal crack will appear as a white line) and can cause confusion. The density range will be severely limited, although within the middle of the density range a contrast very similar to the original film can be achieved. If a compressed contrast-range is acceptable, there are at least five grades of contrast of printing paper, and much can be achieved by local masking and shading. This is the cheapest method of copying.
2. To make a print on paper, through an intermediate negative. This will result in the same type of image as on the original (a crack will be a dark line). The intermediate negative should be made on a slow, moderate-contrast photographic film, and a good and economic technique is to make the intermediate through a copy-

ing camera, with the radiograph on an illuminated screen. This method, being two-stage, gives more scope for control and tone manipulation.

3. Special duplicating film is available, on which the original radiograph is contact-printed, and which on normal development produces the same density tones as the original. By using a mixture of ultra-violet and white light for printing the final contrast can be controlled to a limited extent. If the density range of the original is limited and the average density is low, a duplicate radiograph very close to the original in contrast and range can be obtained, but the duplicate is on a single-emulsion film, which is not capable of high densities or of high contrasts at high densities.

4. For storage purposes radiographs can be copied on to 35 mm or smaller recording film, but the photographic techniques are critical, and such copying film will not cope with the long density range which is present in some radiographs. Also, in copying large radiographs, a size reduction of 10:1 or more is being used, which limits the recordable detail on the original.

5. A radiograph can be copied by using it as a specimen and taking another radiograph of it using very low energy (10–15 kV) X-rays. The varying amounts of silver in different parts of the image provide differences in absorption proportional to the image densities. The result is usually a lower contrast copy than the original film, but the method has some special advantages if the original is of very high density, say greater than four.

6. A radiograph can be imaged by a CCTV camera and the signal digitized and stored on optical disc. Using a 512×512 pixel raster the time-to-store is a few seconds per picture, and there is no permanent consumption of storage material. The apparatus required is expensive, and again there is the limitation that the CCTV camera, having a limited dynamic range, will handle only a limited film density at one setting; to record a long density-range radiograph may therefore require several copies, each covering part of the total density range.

Some insurance organizations' long-term archive requirements for radiographs of their insured items, for example, pressure vessel welds, have created problems in storage space and in the amount of photographic silver held in such a store; pressure for reduced size, or non-photographic copies is therefore considerable. Xerox copying processes, 35 mm film or smaller and microfiche copies are all possible, but only at considerable loss in image quality compared with the original radiograph, unless this has only a limited density range.

REFERENCES

[1] Ruault, P. A. (1969) *Proc. 5th Int. Symp. Industrial Radiography*, Antwerp, Belgium, Agfa-Gevaert.

[2] International Institute of Welding (1969) *Welding in the World*, **7**(4), 200.

[3] Redmayne, I. (1979) *Brit. J. NDT*, **21**(5), 275.

[4] Watanabe, T. (1979) *Brit. J. NDT*, **21**(5), 299.

[5] Glaisher, W. H., Betteridge, W. and Eboral, R. (1944) *J. Inst. Metals*, **70**, 81.

[6] van Horn, K. R. (1958) *Non-destructive Testing (USA)*, **16**(2), 176.

[7] Halmshaw, R. (1965) *Brit. J. NDT*, **7**(3), 63.

[8] Sharpe, R. S. (1966) *Brit. J. NDT*, **8**(1), 17.

[9] DuPont Imaging Systems, Wilmington, USA (1993) technical literature.

[10] Advanced Video Products, Littleton, USA (1994) technical literature.

[11] Forli, O. and Pettersen, B. (1987) *Proc. 4th Euro. Conf. NDT*, London.

[12] Forli, O. and Pettersen, B. (1991) *IIW/IIIS Report IIW-V-969*.

[13] Halmshaw, R. (1988) Capabilities and limitations of NDT in *Radiographic Methods*, British Institute of NDT, Chapter 3.

[14] Halmshaw, R. (1994) *Brit. J. NDT*, **36**(3), 146.

[15] Todd-Pokropek, A. (1981) in *Physical Aspects of Medical Imaging* (eds B. M. Moores, R. P. Parker, and B. R. Pullan), Wiley, Chapter 5.

10

Safety problems in radiography: units of radiation

10.1 INTRODUCTION

The energy absorbed from ionizing radiation when it is incident on living tissue can result in damage or destruction of the cells. The principal reactions are burns, dermatitis, cancer induction and blood changes, and there may be direct damage to chromosomes in individual cells which are responsible for reproduction. It is necessary, therefore, to consider both radiation-induced disease in the individual and genetic damage to part of the total population, which may affect future generations. Both types of damage are considered in fixing internationally-agreed dose limitations. At present it is assumed that there is no wholly safe dose of ionizing radiation, so the risk assumed by an individual or a population group must be balanced against the benefits derived from the uses of X-rays etc. An attempt is also made to ensure that the genetic consequence of the agreed dose limit, in the foreseeable future, has an acceptable limit. The International Committee for Radiological Protection (IRCP) is the international body from which recommendations on protection originate, and these are endorsed by national committees and made law in individual countries or by bodies such as the EC, sometimes with minor changes in the detail of recommended procedures. The international regulations are periodically updated [1–3].

10.2 RADIATION SOURCES

The background level of ionizing radiation is not zero. There is a measurable intensity arising from cosmic rays, that is, from high-energy atomic particles entering the earth's atmosphere from outer space and liberating ionizing radiation on absorption. There are radioactive elements occurring naturally in the earth as minerals. Some common elements have radioactive isotopes which occur naturally in small traces; there are minute traces of radioactive potassium in all naturally-occurring potassium compounds, including those in the human body. The radio-activity in some common rocks, notably granite, makes the natural background of ionizing radiation considerably higher in some cities than others. Some industrial processes release small amounts of radio-activity. Since the first atomic bomb, radioactive substances have been released into the atmosphere and some of these have a long half-life and contribute to the background ionizing radiation intensity. Some of these radioactive substances have been absorbed by vegeta-tion, get into the food chain, and are eventually absorbed into the human body. Finally, ionizing radiation is used by the medical profes-sion for diagnostic purposes and the doses of radiation given to indivi-duals can be large. In addition, in mass surveys such as chest radiography, large sections of the population each received a very small dose of radiation.

For comparison the following values provide an appreciation of the magnitudes of ionizing radiations.

- external exposure to naturally-occurring radioactive isotopes and cosmic radiation: approximately 600 μSv per year (0.06 rem per year);
- total background radiation, external and internal: between 1 and 5 mSv per year (0.1–0.5 rem per year) depending on location (1.86 mSv per year is taken as the figure for London);
- fatal whole body dose of radiation (short period dose): 2–8 Sv (200–800 rem);
- dose to produce density 2 on radiographic film: 2–20 mSv (0.2–2 rem);
- output of typical 250 kV X-ray set: 30 R min^{-1} at 1 m;
- output of large linac: 6000 R min^{-1} at 1 m;
- output of 10 Ci cobalt-60 gamma-ray source: 0.2 R min^{-1} at 1 m.

10.3 RADIATION UNITS

The units of radiation used to measure dose and dose-rate are quite complicated and have been made further so by a 1975 ICRP recommen-dation [4] that a new series of units should be used, to be more com-patible with the SI system. A further report on SI units in radiation protection was issued in 1985 [5]. These new units are not yet in wide-spread use in industrial radiology. In the older better-known units the quantity or exposure-dose of ionizing radiation is measured in roent-gens, defined as the amount of radiation that produces in 1 cm^3 of dry air at 0°C and 760 mm Hg pressure (i.e. 0.001 293 g of air) ions carrying 1 esu of charge of either sign. The roentgen is thus a measure of the amount of energy absorbed by a given volume of air from a beam of radiation, and corresponds to 83.3 erg g^{-1}. In personnel dosage work the milliroentgen is often used, and outputs of radiation sources are given as dose-rates in R min^{-1} or R h^{-1} at a distance of 1 m (Rmm or Rhm). In the SI system the roentgen is not essential and the SI unit for exposure is the coulomb per kilogram (C kg^{-1}) which is equal to 3876 R. A unit of absorbed dose, the rad, was established by the ICRU in 1954 and is defined as the amount of any radiation that results in the absorption of 100 erg g^{-1} in any material. So far as absorption in air is concerned, the roentgen and rad differ only slightly in magnitude and the ICRU has agreed that the two units can be interchangeable so far as personnel protection measurements are concerned for radiation energies less than 3 MeV. It is necessary to remember, however, that the roentgen is a unit of exposure and the rad of absorbed dose, and that these are different concepts. In 1975 the ICRP recommended that a new unit of absorbed dose be brought into use, the gray: 1 Gy = 100 rads = 1 J/kg^{-1}; conversely 1 R = 8.732 × 10^{-3} Gy. An alternative term sometimes used for absorbed dose is 'kerma' (derved from 'kinetic energy released in media'). Kerma is used to describe the transfer of energy from photons to electrons.

The third unit, the rem (roentgen-equivalent-man), is a complicated unit that takes account of the product of the energy absorption and its biological effect, and is the absorbed dose of any ionizing radiation which has the same biological effect as 1 rad of X-radiation in terms of absorbed dose in air [6]. For medium energy X-rays the rem and rad can be taken to be numerically equivalent, but for very high energy radiation and for other ionizing radiations or particles, there is a conversion factor. In 1975 the ICRP also recommended that the unit of dose-equivalent be renamed the sievert (Sv) and that 1 Sv = 100 rem.

For clarity in this chapter, measurements will be given in both new and old units, as far as possible, but for the output of radiographic equipment everyone still uses roentgens per minute at 1 m (Rmm or Rhm, C kg min^{-1} at 1 m is never quoted).

These definitions do not make accurate measurements of these quantities easy to make as a 'free-air' ionization chamber, in which the effects of the wall material enclosing the air-space are negligible, is difficult to design. The walls must be thin enough not to cause appreciable absorption, and corrections must be made for photoelectrons emitted into the enclosed air-space from the walls. Photographic film can be used to measure radiation by using a calibration curve of film density against dose, but the response of film is radiation energy dependent, so that measurements on a heterogeneous beam are subject to error.

10.4 PERMISSIBLE DOSE LIMITS

Regulations on maximum permissible dose and dose-rates are comparatively complex. In 1921 the British Radiological Protection Board recommended a 'tolerance dose' of 1 R per week and it was generally thought that below this level ionizing radiation produced no measurable effect. In 1928 the ICRP was set up and over later years issued a series of directives. In 1954 the IRCP replaced the concept of tolerance dose by 'maximum permissible dose' (MPD) and in 1958 proposed an MPD of 0.3 rem per week (3 mGy per week) for critical human organs. In 1977 the ICRP produced Publication No. 26 in which the concept of MPD disappeared and ALARA ('as low as reasonably achievable') was introduced.

It had already been realized that radiation dose should be divided into somatic (the effect on the person irradiated) and genetic (the longer-term effect on descendants and the population at large). Using available data for natural background dose-levels, Publication 26 proposed dose-equivalent limits of:

- whole body dose for radiation workers: 50 mSv per year (5 rem per year).
- whole body dose for the general public: 5 mSv per year (0.5 rem per year).

Other figures were proposed for special cases (gonads, eye lens, pregnant women, skin, hands etc.).

The concept of 'quality factor' (QF) was also introduced, recognizing that neutrons, alpha-particles etc., could be more lethal. For X- and gamma-rays up to 3 MeV, QF = 1; for neutrons, QF is up to 10. In earlier documents workers were divided into two groups:

1. classified workers: male adults over 18 years of age who are employed on radiation work, who carry monitoring films and have periodic medical examinations;
2. persons other than classified workers; the rest of the population.

But in Publication 26 the term 'designated person' was used in place of 'classified worker' and these personnel are described as anyone whose dose might exceed 3/10ths of the annual permitted dose of 50 mSv.

In 1989 the Health & Safety Executive (UK) produced a report [3] stating that National Radiological Protection Board (NRPB) advice was to reduce the 50 mSv per year to 15 mSv per year (1.5 rem per year) and to reduce the dose to the public to 1 mSv per year (0.1 rem per year), but these new figures have not, to date, been implemented. The regulations state that the dose record of an employee must be checked once every three months.

The *Ionising Regulations* [1] cover many other points such as the appointment and duties of a Radiation Protection Officer, Radiation Safety Officer and Supervisory Medical Officer (the reader is referred to the full documents for details [1–3]. A designated (classified) worker must carry radiation measuring devices of some form (section 10.5) but need not have regular medical supervision unless there is evidence of dosage above the recommended limits. A later document (Health & Safety Executive (UK), No. 2379:1993) provides further details on the working requirements for 'outside' workers, i.e. for site work [7].

The dose of 50 mSv per year is usually converted to 1 mSv per week or $25 \mu Sv\ h^{-1}$ (2.5 mrem h^{-1}) and this is the dose-rate to be applied to all areas surrounding radiographic installations which are accessible to only classified workers. In the code of practice produced by the EC, some account is also taken of the rate of usage of X-ray equipment, that is, it is not necessary to assume that the X-ray set is generating X-rays for 40 h per working week.

For radiographic installations where non-designated personnel might-have access to the vicinity, the ICRP and the EC directives require that the dose-rate 'averaged over a period considered appropriate by the Radiation Protection Officer' shall not exceed $7.5 \mu Sv\ h^{-1}$, (0.75 mrem h^{-1}) so that this is the radiation level which must not be exceeded on the outside of most radiographic installations. This is called a 'controlled area'. However, it is not necessary to designate places as controlled areas if there are effective engineering controls which prevent any person from being exposed to an instantaneous dose-rate exceeding $7.5 \mu Sv\ h^{-1}$.

There appear to be three main groups of problem concerned with radiation protection in industrial radiography. It is assumed that un-sealed radioactive sources are not used. The first problem is the design of a radiographic installation in which there are interlocking doors such that X-rays or gamma-rays cannot be switched on until the room (laboratory) is empty of personnel and the doors closed. The aim should be to achieve a dose-rate of less than $7.5 \mu Sv\ h^{-1}$ on the outside walls, wherever the equipment inside is located or oriented. As already mentioned, the Radiation Protection Officer may be able to recommend an occupancy factor, for example, that no personnel will be in the vicinity of the walls for longer than 10 h per week, in which case the officer may allow a relaxation on the dose-rate figure. Similarly, if the equipment is to be activated for only a limited period, a second allowance might be agreed. It need not be assumed that the X-ray set is switched on at full power for 40 h per week. These are important points in the cost of shielding, particularly when high energy X-rays are being considered.

The second problem is the use of X- and gamma-rays for site work, i.e. on a temporary set-up. The main problem here is the exclusion of non-classified personnel from any area where the dose-rate may be greater than 7.5 µSv h⁻¹, by barriers or local shielding. The responsibilities on site for 'outside workers' involved in radiography have recently been expressed in further detail [8]. Much more emphasis is laid on prior notification of site work and in keeping the need for site-radiography to a minimum. For underwater radiography by a diver, barriers are not necessary (see section 7 in [2]). For the purposes of starting or terminating exposures, the instantaneous dose-rate should not exceed 2 mSv h⁻¹. There are detailed instructions on warning signals, prior notification of work etc. [2, 3, 7].

The third problem concerns the design and transport of gamma-ray source containers. In general, in the UK an exposure container should comply with BS:5650:1978 (ISO:3999:1977). More details of the requirements are given in section 10.8.

10.5 RADIATION MONITORING EQUIPMENT

Various types of dosemeter, dose-rate meter and film monitoring badges are used.

In the UK all classified workers must wear a radiation badge issued by an approved authority. Originally these were film badges of a standard design containing a photographic (dental) film, which was periodically sent for processing and replaced. The density on the processed film provided a record of the total dose of radiation to which the film (and therefore the wearer of the badge) had been exposed. The advantages of the film-type badge are that it responds to all hazardous external radiations except fast neutrons (for which a separate badge is required) and gives a good indication of the type and energy of the radiation by having a series of filters over different parts of the film, of different thicknesses and materials (Fig. 10.1). It has a good latitude (10 mrad–1000 rad, i.e. 100 mGy to 10 Gy), because of the double-sided

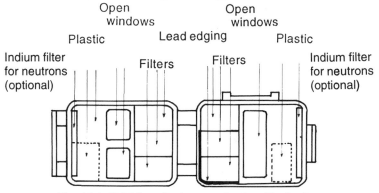

Fig. 10.1 British radiation-monitoring film badge.

film, and provides a permanent record. It is not cheap, and since films are normally worn for four weeks, it does not give a prompt warning of over-exposure.

The processing of film badges is critical and is usually done by a central authority, but in general, in the UK film badges have been preferred to personal ionization chambers. More details of film badges are given in BS:3664:1963 and BS:3890:1965.

Recently, thermoluminescent (TL) personnel badges have begun to replace film badges. TL badges which conform to Euratom specifications consist of an anodized aluminium plate carrying two discs of different thicknesses of PTFE loaded with TL lithium fluoride or lithium borate powder. These plates are sealed in a light-proof cover which is printed with an identification in binary form, and the dosemeter is worn in a plastic holder. On return to the NRPB the dosemeter is processed automatically. The lithium fluoride discs emit light when heated and the amount of the light output is a measure of the absorbed dose of ionizing radiation. The badges are re-usable after a thermal anneal and can be used up to 100 times. The badges can measure doses smaller or larger by a factor of ten than those readable from a film badge. Lithium fluoride is close to tissue in its atomic number, so its response is independent of photon energy. At room temperatures the stored signal is very stable. The disadvantages of the TL badge are the cost and complexity of the processor, and the impossibility of re-reading if the initial reading is not properly recorded, although this latter limitation is not entirely true for doses in excess of 15 mGy. It also gives less information about the nature and energy of the radiation unless combinations of TL materials are used. The advantages are automatic processing so that the doses can easily be transferred to a computer, lower sensitivity to environmental conditions such as humidity and high temperature, and automatic registration and record-keeping. The dosemeters are widely used in the USA, Japan and several European countries. TL dosemeters do not have to be handled in the dark, but are sensitive to ultra-violet. They should not be handled except with forceps, as skin-grease can affect the glow curve during read-out.

For immediate reading, ionization chamber instruments are commonly used, but there are also personnel dosemeters of the quartz fibre or condenser type which can be carried on the person and read periodically. Battery-operated portable alarms are also used; these give an audible warning when a pre-determined dose has been reached and usually consist of Geiger-Müller counter tubes. The best of these instruments are direct-reading, i.e. they can be inspected at any time, to give the integrated dose; some also give an alarm of either dose-rate or pre-set integrated dose, but generally none give a permanent record and they have to be backed up with a film badge.

With all these instruments the response to radiation of different energies is not independent of energy, and the instruments must be calibrated. There is therefore some inaccuracy of measurement when the beam to be measured contains an unknown mixture of energies, such as hard primary radiation together with softer scattered radiation. There are also particular difficulties in accurately monitoring very low X-ray energies.

10.6 PROTECTION DATA

The basic methods of reducing radiation dose-rates are adjustment of distance, time and shielding. To calculate the necessary thicknesses of protective walls it is necessary to have the following data:

1. the radiation output of the sources of radiation (in R min^{-1})
2. the distance between the source and the point of measurement (the dose-rate varies inversely as the square of this distance);
3. the absorption characteristics of the material of the protective walls (in general, as the area of wall irradiated will be large, broad-beam rather than narrow-beam attenuation characteristics are required).

It may be possible to restrict the direct beam to certain areas or directions, in which case some walls need only be thick enough to provide protection against scattered radiation.

For equipment operating below 200 kV the cost of protecting all the walls up to direct-beam standards is unlikely to be large, but for equipment over 300 kV the cost of protective walls can be very large and if an extra small, local barrier can provide all the direct-beam protection which is needed by restricting the possible beam directions, this may be economically important. Scattered radiation due to the Compton effect is on average very much less penetrating than the direct beam and consequently thinner barriers are adequate outside the main X-ray beam.

Protective materials

For medium-energy X-rays, lead is the most suitable shielding material, but concrete is also used. Concrete and brick will generate much more scattered radiation and should be covered with a metal sheet.

For X-ray energies below the megavoltage region, lead is a much more efficient protection material than low-density building materials, but in practice a combination of the two is usually used. Table 10.1 gives equivalent thickness of lead, steel, brick and concrete for equal absorption. The values of these equivalents are necessarily approximate. The exact values depend on how much scattered radiation is included in the transmitted intensity and also whether one is considering a thin or thick

Table 10.1 Corresponding thicknesses for the same absorption (equivalent thicknesses in mm)

	Lead	*Steel*	*Brick* (density = 1.6)	*Concrete* (density = 2.4 g cm^{-3})
100 kV X-rays	1	17	120	75
200 kV X-rays	1	14	110	65
300 kV X-rays	1	8	60	25
400 kV X-rays	1	7	45	18.5
1 MV X-rays	1	4	–	6.2
5 MV X-rays	1	1.5	–	5
20 MV X-rays	1	1.5	–	4.7
Iridium-192 gamma-rays	1	4	–	8.8
Cobalt-60 gamma-rays	1	1.6	–	5.4

barrier. In practice the approximate values will give a guide to the designers.

For high energy radiation, it will be seen from Table 10.1 that the thicknesses of materials required for equal absorption are directly related to their physical density, that is, the protection is proportional to the weight per unit area of the shielding. It is economically sound to use the cheapest building material, and this is usually concrete. Besides plain concrete, special high-density concrete made from barium sulphate (two parts barium sulphate, two parts sand, one part cement) is sometimes used; lead–concrete and concrete containing 50% iron oxide have also been marketed. These materials have much higher physical densities than ordinary concrete, and so provide the same protection with smaller thicknesses, but their use depends largely on their cost.

For small thicknesses of lead, up to 6 mm, a lead-ply material in which one layer of the plywood is replaced by a lead sheet bonded on to the wood layers is convenient to handle. For thicker lead walls interlocking lead bricks, which have four vee-shaped faces and two flat faces, are available. Opposite faces have external and re-entrant vees, so the bricks interlock to build a parallel-faced wall with no straight-through gaps, without the need for mortar. These bricks are the subject of British Standard BS 4513, 1969. Similar interlocking bricks are available in 5 in thick heavy concrete.

10.7 CALCULATION OF PROTECTIVE BARRIER THICKNESSES

The maximum radiation output of any source of ionizing radiation should be measured directly, but as a guide Table 10.2 shows typical outputs of present-day equipment. Figures 10.2–10.5 show broad-beam absorption data for various X- and gamma-ray sources for lead and for standard concrete (see also BS 4094, 1971 Parts I and II). Table 10.3 gives

Table 10.2 Typical radiation outputs

Source	Radiation output (Rmin^{-1} 1 m, unfiltered beam)
100 kV X-rays; 5 mA	1
400 kV X-rays; 10 mA	50
8 MV X-rays (linac)	6000
20 Ci Ir-192 gamma-rays	0.16
10 Ci Co-60 gamma-rays	0.22

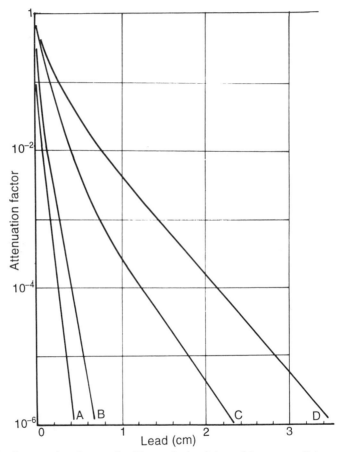

Fig. 10.2 Attenuation factors for X-rays in lead, broad-beam conditions. Curve A = 100 kV X-rays; Curve B = 200 kV X-rays; Curve C = 300 kV X-rays; Curve D = 400 kV X-rays.

values of the final tenth-value thickness in lead and concrete for various radiations; this is the thickness of absorbing material which reduces the

Table 10.3   Tenth-value thicknesses

Radiation	Lead (mm)	Concrete (mm)
100 kV X-rays	0.9	55
250 kV X-rays	3	90
400 kV X-rays	8	100
1 MV X-rays	25	150
Ir-192 gamma-rays	20	142
Cs-137 gamma-rays	23	169
Co-60 gamma-rays	40	220

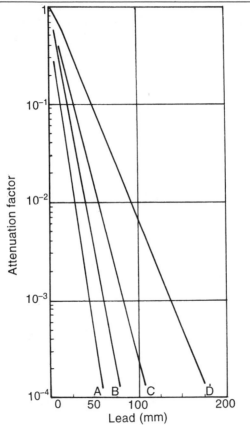

Fig. 10.3 Attenuation of gamma-rays in lead, broad-beam conditions. Curve A = Ir-192; Curve B = Cs-137; Curve C = Cs-134; Curve D = Co-60.

radiation intensity to 1/10, and represents the slope of the absorption curve at large thicknesses.

For example, consider a 400 kV, 10 mA X-ray set with an output 50 R min^{-1} at 1 m and determine the thickness of concrete needed at 3 m distance for the dose rate to be reduced to 2.5 mR h^{-1}. By the inverse square law, the dose-rate at 3 m is $50 \times 60/9 = 333$ R h^{-1}. Thus, the attenuation factor needed is

$$\frac{2.5 \times 10^{-3}}{333} = 0.000\,007\,5$$

From Fig. 10.4 this attenuation is provided by 58 cm of concrete. Alternatively, 6 mm lead will provide an attenuation of 0.04 and the additional attenuation will be provided by 43 cm of concrete.

This example takes the worst case of the direct beam, but in practice this beam will usually be directed at the floor or at a supplementary barrier, or at the specimen, and most of the wall area has only to provide protection against scattered radiation. The intensity of scattered radi-

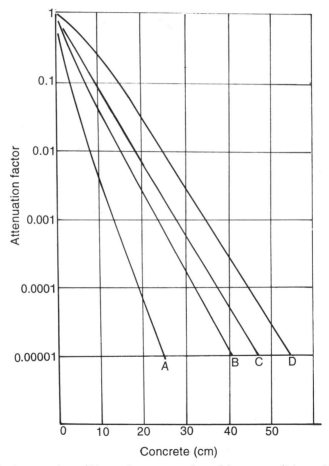

Fig. 10.4 Attenuation of X-rays in concrete, broad-beam conditions. Curve A = 100 kV X-rays; Curve B = 200 kV X-rays; Curve C = 300 kV X-rays; Curve D = 400 kV X-rays.

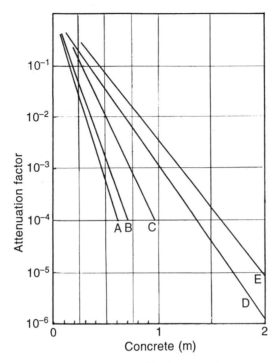

Fig. 10.5 Attenuation of gamma-rays and high energy X-rays in concrete, broad-beam conditions. Curve A = Ir-192; Curve B = Cs-137; Curve C = Co-60; Curve D = 6 MV X-rays; Curve E = 10 MV X-rays.

ation depends on the area of the specimen irradiated, the angle of the scatter, and the distances involved; it is not easy to calculate with any accuracy. As a guide, it is unlikely that the scattered radiation intensity will be greater than 0.1% of the primary beam, and the mean energy of the scattered radiation may be taken to be about 300 keV maximum for high-energy equipment. Table 10.4 gives guidance on the relative thicknesses when only scattered radiation is involved, from data by Balteau [9].

Additional data on the calculation of barrier thicknesses for protection against secondary radiation are given in BS 4094, 1971 Parts I and II.

Table 10.4 Protection requirements at 4 m distance for 0.5 rem per year

Equipment	Direct and scattered radiation		Scattered radiation only	
	Lead (mm)	*Concrete* (mm)	*Lead* (mm)	*Concrete* (mm)
100 kV, 5 mA	2.7	200	0.8	60
200 kV, 10 mA	6.7	435	2.4	200
300 kV, 10 mA	21.4	540	6.9	280
400 kV, 10 mA	37.2	615	13.9	345

10.8 GAMMA-RAY SOURCE CONTAINERS

Regulations governing the design of gamma-ray source containers are given in BS 5650, 1978, which is essentially ISO-3999-1977.

Exposure containers are divided into two categories:

1. the source is not removed for the exposure to be made;
2. the source is projected out for exposure (cable, pneumatic etc.).

The removal of a source on a handling rod or other device, which used to be common practice, is now banned in some countries and is not discussed in these standards. Containers are further sub-divided into: P, one-man portable; M, mobile; and F, fixed. The exposure rate limits for these divisions are shown in Table 10.5.

Tests are detailed for vibration, shock, endurance, and two types of accidental drop. For the Category 2 containers, there are kinking, crushing and tensile tests for the projection sheath (the tube through which the source is moved to the exposure position). There are additional details on identification markings and signals on the container for open and shut positions. Most containers are made of either lead in a steel or brass shell, depleted uranium, or tungsten alloy, although there is some doubt whether a lead-in-steel container will satisfy the thermal tests required for road transport, unless it is provided with additional thermal insulation.

The calculation of container thicknesses is very similar to the examples already given. Table 10.6 gives half-value thicknesses for the two most-commonly used radioisotopes for the materials used in container construction.

Table 10.5 Gamma-ray source containers: exposure-rate limits (mR h^{-1})

Type	External surface	50 mm *from* surface	1 m *from* surface
P	200	50	2
M	200	100	5
F	200	100	10

Table 10.6 Half-value thicknesses

Material	Gamma-ray isotope	Second HVT (mm)	Notes
Steel	Ir-192	9.2	Density 7.8 g cm^{-3}
Lead	Ir-192	2.8	Density 11.3 g cm^{-3}
Tungsten alloy	Ir-192	2.4	Density 17.7 g cm^{-3}
Uranium	Ir-192	2.2	Density 18.8 g cm^{-3}
Steel	Co-60	16.4	–
Lead	Co-60	10.6	–
Tungsten alloy	Co-60	6.92	–
Uranium	Co-60	5.67	–

Transport of gamma-ray sources

If the exposure container is also to be used as the container in which the source is transported on public roads, either with or without additional protection, as is fairly common practice nowadays, it needs to satisfy additional regulations. The International Atomic Energy Authority has categorized two types of package [10]; type A is for very small quantities of radionuclide; type B will apply to virtually all sources used in industrial radiography, and a type B package must be designed to sustain any credible accident during transport. This has been interpreted to mean, for road transport, a collision with an oil tanker and subsequent fire, and a container must be able to withstand a drop test followed by a 30 min fire at 1000 °C with no escape of the radioactive contents. If the container has to cross international boundaries there are complex notification procedures to 'competent authorities'.

Package labelling is also the subject of international recommendations and one must refer to the national standards for the detailed regulations on most of these points.

10.9 GENERAL SAFETY REQUIREMENTS FOR RADIOGRAPHIC LABORATORIES

Most national regulations on radiological protection require a number of additional safety requirements to be met. The rigorousness with which these are applied depends on individual safety inspectorates, common sense and local prejudice. The ALARA principle tends to become ALARP ('as low as reasonably practical'). Obviously it is not necessary to be as strict over alarms etc. for a small cubicle containing a 100 kV X-ray set as for a large laboratory containing a powerful linac; some authorities will accept electrical interlock systems, others insist on mechanical systems.

The main points are summarized below, but it is emphasized that for full details of the necessary procedures, particularly the certification required, the formal legal documents and the codes of practice of each individual country must be studied.

1. If any door of a radiographic enclosure can be opened, means must be provided so that the equipment is automatically switched off and cannot be switched on while the door is open. With an X-ray set this can be done either with interlocking electrical switches or by mechanical means. A widely used system is the Castell key in which a key to operate the X-ray machine cannot be obtained from an exchange box until the keys locking the room doors are inserted; in turn, the door keys cannot be obtained until the doors are properly closed and locked. The interchange system works in reverse and can only be

overridden by the use of a special master key. With a gamma-ray source it is possible to interlock the container-opening mechanism to the door-locking system, but most workers regard a gamma-ray source as portable equipment and do not wish to have this limitation on movement.

2. In a large radiographic enclosure there should be an emergency exit through which persons can leave without delay, and it is mandatory to provide an emergency switch inside the enclosure which will switch off the equipment.

3. Audible warnings or visible lights are required to give warning that equipment is about to be energized or a source exposed.

4. There should be a separate warning light to show when the source is emitting radiation. It is good practice to duplicate this light in all places where workers may have access around the enclosure, and it is quite straightforward to connect such lights into the high tension (HT) switch of an X-ray set.

5. Suitable warning notices of ionizing radiation are also required.

6. With large laboratories where there may be 'dead' areas, CCTV cameras with monitors in the control room can be used. An inter-communication system is also desirable.

For site work, some of the above provisions cannot always be applied and conditions vary so much that only general recommendations are possible [5]. Common sense, radiation monitoring and direct surveillance will usually provide safe working conditions without the elaboration of interlocks etc. required on a permanent enclosure.

Safety monitoring

On any radiation enclosure, however detailed the design, a radiation survey must be undertaken before use, or after any alterations. This should be done with the source operating at its maximum output and pointing in all directions in which it is likely to be used. If there are restrictions in the permissible beam directions which are not physically controlled by mechanical or electrical limiting devices, permanent notices explaining these are necessary. Some X-ray sets can continue to emit X-rays for a few seconds after the HT has been switched off if the filament current is left on, due to the capacitance of the HT circuit which can continue to provide a rapidly-decaying voltage across the tube. With most X-ray enclosures this effect is never significant because of the time required to enter the enclosure, but with an X-ray cabinet with a small door there is a possibility of opening this so quickly that the radiation level has not dropped quite to zero. If necessary, an automatic delay can be introduced.

With very high energy (megavoltage) equipment, short-life radioactivity can be induced in some materials by photonuclear reaction, and after long exposures it is desirable to monitor the level of any activity before handling the specimens. The energy thresholds for inducing radioactivity are, for example 5.0 MeV for iron, 6.1 MeV for aluminium, 11 MeV for copper and 1.8 MeV for phosphorus.

'Sky shine'

High energy X-ray equipment, particularly in the megavoltage range, produces side-lobes of radiation outside the main direct beam, and as this is radiation with a large penetrating power it can travel large distances in air. In some machines this unwanted radiation is absorbed close to the target, but in many machines it can travel upwards and outwards, and if the laboratory has a relatively thin roof it can spread outside the laboratory. It is not usually feasible to build a roof of the same thickness as the sidewalls of the building, so that account must be taken of the radiation extending into the air above the laboratory, which is absorbed and scattered back to regions outside the laboratory at ground level (Fig. 10.6), the so-called 'sky-shine'.

Calculation of the intensity of sky shine is complex, requiring a knowledge of the angular distribution of the X-ray output, the scattering coefficient at different angles and the coefficient of sky shine (defined as a function of angle of scatter and source distance). It is usually easier to make measurements on site.

Calculations [11] have shown that the sky shine is at a maximum at a distance from the building wall approximately equal to the wall height (Fig. 10.6); it was also shown that with an X-ray output of 1000 R min^{-1} at

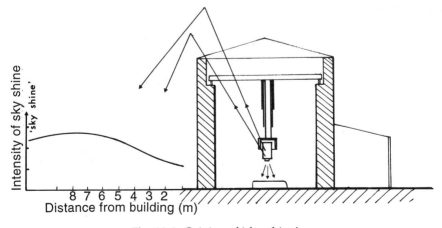

Fig. 10.6 Origins of 'sky-shine'.

1 m, the level of sky shine is around $2\,\text{mR}\,\text{h}^{-1}$ for a present-day linac (8 MeV) even when there is no absorption in the roof. If space is available, therefore, it may be less expensive to exclude personnel from around the X-ray laboratory by fences, etc. than to provide a thick roof. Additional data on the calculation of sky shine dose-rates, including correction factors for barrier wall heights, are given in BS 4094, 1971, Part 1.

REFERENCES

[1] HMSO (1985) *The Ionising Radiations Regulations No. 1333*.
[2] Health and Safety Commission (1985). *The Protection of Persons against Ionising Radiations arising from any work activity. Approved Code of Practice*, HMSO, London.
[3] Health and Safety Commission, Working Group on Ionising Radiations. (1989) *1st Progress Report 1986–87*.
[4] International Commission on Radiation Protection (1975) *Radiology*, **125**(2), 492.
[5] UK National Commission on Radiological Protection (1985) *Report No. 82*.
[6] International Commission on Radiological Protection (1979) *Report No. 31*.
[7] H.M. Factory Inspectorate (1975) *Code of Practice for Site Radiography*, Harrap, London.
[8] HMSO (1993) *Publication No. 2379*.
[9] Balteau and Cie, Liège, Belgium (1971) *Data Sheet on Protection*.
[10] International Atomic Energy Authority (1973) *Regulations for Safe Transport of Radioactive Materials*.
[11] Fujita, H. (1973) *ARS News*, **7**(6), 1.

11

Fluoroscopy, image intensifiers, television systems and tomography

11.1 INTRODUCTION

The direct production of a visible image on a fluorescent screen, X-ray fluoroscopy, has been known and used since the early days of X-rays but now has few applications in industrial radiography. In fluoroscopy the X-radiation that is transmitted through the specimen falls on to a screen which fluoresces, that is, emits light within the visible part of the spectrum; a visible image is produced on the fluorescent screen due to differential absorption in the different thicknesses of the specimen. The thinner, less absorbent parts of the specimen are seen as brighter areas on the screen, so that the tonal range is reversed compared with a film radiograph seen on an illuminated film-viewing screen. Cavities are brighter, not darker. The potentialities of 'real-time' imaging, particularly in medical diagnostic radiology, and the ability to see moving images, have led to major developments, e.g. image intensifiers and CCTV-fluoroscopic systems.

Industrially, fluoroscopy has the advantage over film radiography that an image is obtained without the use of consumable recording material. The major disadvantage is that the detail and sensitivity observable are usually much poorer. There are three main reasons for this poorer sensitivity.

1. A typical fluoroscopic screen image is very dim (typically $0.3 - 0.003\,\mathrm{cd\,m^{-2}}$). At these low luminances the human eye, even when fully dark-adapted, is incapable of perceiving such small contrasts or such fine detail as can be discerned on a film radiograph.
2. Fluoroscopic screens are generally constructed to give the brightest possible image and, compared with film, are very coarse-grained and

incapable of resolving fine detail (MTF as low as 25% at 3 lines mm^{-1} is typical, compared with 20–30 lines mm^{-1} on film).

3. Fluoroscopic screens have a contrast gradient of unity, whereas film has a contrast gradient of 4–6 at the densities normally used, so a small difference in X-ray intensity (across, say, the image of a small flaw) is enhanced by this factor in the film recording process.

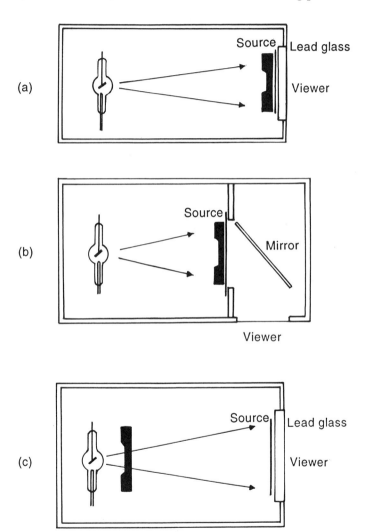

Fig. 11.1 Typical fluoroscopic arrangements. (a) = specimen close to screen, the screen is viewed through a suitable thickness of lead glass; (b) = conventional arrangement with mirror viewing; (c) = enlarged image system with fine-focus X-ray tube.

Conventional direct fluoroscopy is nowadays hardly ever used for flaw detection, but equipments are still built for simple inspection of assemblies, non-metallics etc. The essence of the design of these is to enable the operator to view the fluorescent screen without any radiation hazard, and to insert and manipulate the specimen in safety.

The three basic designs are shown in Fig. 11.1. In Fig. 11.1(a) there is a thick sheet of special lead glass between the operator and screen. This lead glass contains a high proportion of lead salts in its composition and therefore strongly absorbs X-rays. As fluoroscopy seldom uses X-ray energies greater than 150 keV, a 10 mm sheet of glass can be equivalent to 2–2.5 mm of metallic lead. Modern lead glass made for this purpose has a good transparency to light and it is quite practical to use 30 mm or more of glass. Figure 11.1(b) shows an alternative method using a mirror to view the fluorescent screen; multiple mirrors can also be used. For this method a mirror of approximately the same size as the screen is needed, and it should be surface-silvered to eliminate double images. In both these methods the specimen is close to the screen and X-ray tubes with large outputs are desirable to obtain as bright an image as possible. Figure 11.1(c) shows an alternative method, using an X-ray tube with a small focal spot size and projective magnification: an enlarged image of the specimen is obtained on the screen, which in some circumstances enables more detail to be seen because of its larger size [1].

Fluoroscopic equipment therefore basically consists of a lead-lined cabinet containing the X-ray set, mechanisms for supporting and moving the specimen and a fluorescent screen. Since the image on the screen is usually rather dim, it is also necessary to have the whole cabinet in a dark room, or to have very efficient hoods over the screen to exclude extraneous light. The image on a fluorescent screen is reversed in tone, so if a small object is placed on the specimen table with no masking, the surrounding area will be very bright. This brightness can obscure detail in the image of the specimen. Masking of edges and thin sections of the specimen is therefore essential; a method which is sometimes adequate is to use a thin metal sheet between the specimen and screen.

11.2 IMAGE INTENSIFIER TUBES

The very brief description of X-ray fluoroscopy showed that the chief limitation is the low brightness of the screen image, and a number of methods have been devised to amplify this brightness. Practically all the development work has been for medical equipment, so the emphasis has been on equipment designed to work at less than 140 kV.

The first developments were electronic X-ray image intensifier tubes [2, 3] in which the image on a fluorescent screen at one end of the tube is converted successively to light, then to electrons, these electrons being focused on to a smaller screen where the image is reconverted to light. All the screens must be inside a vacuum envelope. The process of reduction in size and electron acceleration produces a gain in brightness and the final image can be 300–1000 times brighter than a simple fluorescent screen. The final image is bright enough to view in ordinary room lighting, but direct viewing is rarely used and the normal method of viewing nowadays is through a CCTV system.

The main development in X-ray image intensifier tubes in the last 30 years has been the construction of primary conversion screens which absorb more of the incident X-rays. In modern tubes this screen usually consists of columnar crystals of caesium iodide, CsI(Na), with the columnar axis oriented along the direction of the X-ray beam; this screen is usually coated on the inner face of the glass envelope. By this means, a much thicker screen is obtained which absorbs and converts a greater percentage of the X-rays. Most of these tubes are still built for medical use, but a few have been designed with thicker screens, for industrial applications using higher kilovoltages. Apart from the columnar crystal shape, caesium iodide screens can be made with a higher packing density than conventional zinc sulphide screens; also, caesium iodide has a higher absorption in the region from 40–70 keV.

Today, X-ray image intensifier tubes are still widely used, but are in competition with at least two other major methods of producing an X-ray image on a television monitor screen.

11.3 RADIOSCOPIC EQUIPMENT

It is convenient to divide this equipment into three main groups, 1–3. With Group 1 equipment (Fig.11.2) a conversion screen is placed behind the specimen to convert the X-ray image (i.e. the X-rays transmitted through the specimen) to light. This screen is 'open', not in a vacuum tube. The light image on this conversion screen is picked up by a CCTV camera of suitable sensitivity, amplified, converted to digital data (pixels) and held in a framestore from where it can be extracted for image processing and then displayed on a TV monitor. Some CCTV cameras produce digital outputs directly without the use of an A/D converter.

The special point about Group 1 equipment is that the primary screen is open, so it can be of any size of any fluorescent material, and can be interchanged. New materials for the primary screen have been proposed which have a greater conversion efficiency than conventional ZnS:CdS screens [4]. Many of these newer screens are made from commercially-protected materials but screens consisting of single crystals of sheet caesium iodide, up to 300 mm diameter for use with high energy X-rays, were reported from the USSR [5]. The light image on such a screen is not

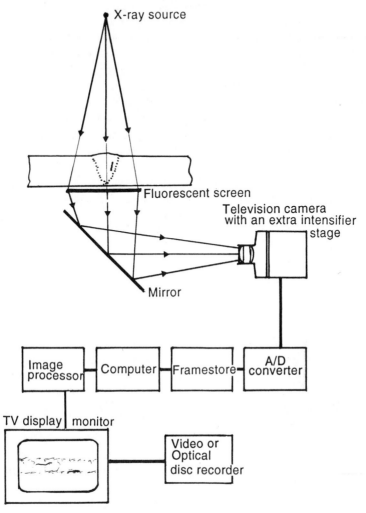

Fig. 11.2 1-type 'open' screen equipment.

very bright, but this can be overcome by the large amplification which is possible through the TV channels. The sideways diffusion of the image in the slice of crystal is quite small and the thinness of the slice minimizes the optical distortion of off-centre areas of the image.

High-density glass scintillators are available in both solid and fibre forms, which are claimed to give higher X-ray attenuation and greater light output with X-rays above 150 kV. In the fibre form, short lengths of glass fibre are fabricated into mosaic screens. The poor resolution of the fabricated screen can be overcome by using a geometrically-enlarged image [6].

On Group 1 equipment, because the image on the primary screen is usually of very low brightness, a sensitive TV camera is needed, usually one with an extra intensification stage such as a silicon-intensifier vidicon (Sitcon) or an Isocon. These are more expensive than standard vidicons and tend to be less robust and reliable. The development of new TV cameras is, however, proceeding very rapidly.

Group 1 equipment, as with most other radioscopic systems using TV, can be used in complete radiation safety with any energy of X-rays or gamma-rays. Their performance is limited by quite different parameters compared with simple, direct fluoroscopic equipment (section 11.5).

The use of a TV circuit brings considerable advantages in that variable magnification can be used to pick out and examine a local region of the specimen; contrast control over a wide range is possible, also image subtraction techniques to enhance local contrast; video-tape recording, storage and playback are well developed, as is laser-spot optical disc recording, which allows still greater storage capacity.

With Group 2 equipment (Fig. 11.3) a conventional X-ray image intensifier tube is used to convert the X-ray image to light, as already described. A CCTV camera is focused on the intensifier tube output screen, and because this screen is much brighter than in Group 1 equipment, a less sensitive CCTV camera can be employed, such as a vidicon or a solid-state camera. Solid-state cameras have a 2-D light-sensitive array of small elements with a digital output, so that a complete camera consists of only a lens and a sensitive array, with no electron beam scanning. As new light-sensitive arrays are developed, with more elements on the array and greater sensitivity, such cameras are being more widely used instead of vidicons.

The output from the camera is amplified, stored, enhanced and displayed, as with Group 1 equipment. Most X-ray image intensifier tubes can provide $\times 2$ or $\times 3$ electronic magnification, which can increase the image resolution from about 4 lines mm^{-1} to 5 lines mm^{-1}. The X-ray image size is obviously limited to the X-ray intensifier screen diameter, which is commonly 150–300 mm, although larger tubes exist.

The main advantage of Group 2 equipment is that the TV cameras generally used are stable, rugged items requiring little maintenance or adjustment. To date, this type of equipment has been most widely used with X-ray kilovoltages in the range 50–300 kV. Owing to the simplicity, robustness and reliability of vidicons and solid-state cameras, Group 2 equipment is still the most widely used for NDT applications. The disadvantage of this type of equipment is that the user is restricted to the sensitive screen built into the tube, both in size and material. Further, in the past image intensifier tubes have deteriorated in performance over time.

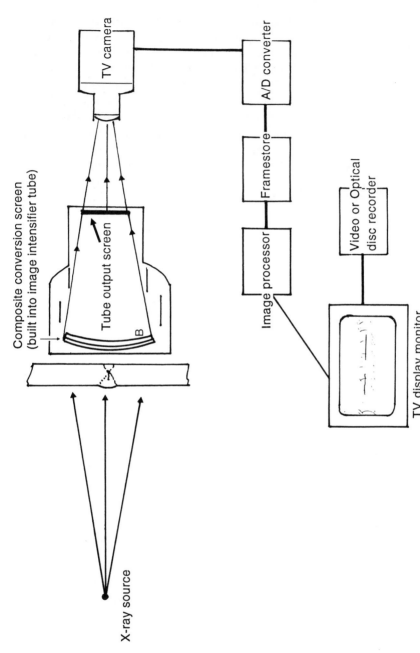

Fig. 11.3 Group 2-type equipment with image intensifier tube.

In 1995, cooled-CCD cameras with sensitivities from 10^{-9}–10^{-11} lux, 2000×2000 pixels (24×24 μm pixels on a 50×50 mm sensitive surface) and a dynamic range of $10^5 : 1$, arrived the market [7]. Pixel sizes as small as 9×9 μm are also offered. The sensitive area is cooled by in-built semiconductors using the thermoelectric Peltier effect, and this reduces the amount of thermally-generated dark current. Such cameras have sufficient sensitivity to be used with Group 1 equipment, as well as with Group 2.

Group 3 equipment (Fig. 11.4) has a linear array of up to 1000 X-ray sensitive elements, each of the order of 1 mm or less in width, which can be constructed using either a piece of fluorescent screen covering a photodiode or a series of charge-coupled devices. If this array is scanned across the specimen and the output of each element taken into a framestore, each position of the linear array is equivalent to one row of pixels. If several hundred rows are recorded and stored as the array is moved,

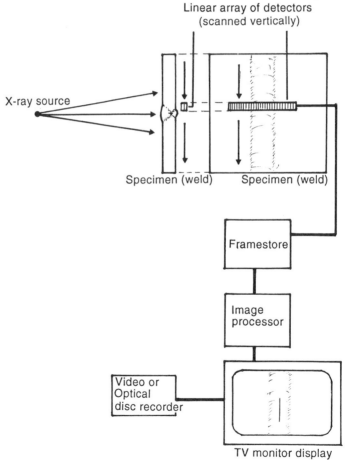

Fig. 11.4 Group 3-type equipment; a scanning linear array system.

the equivalent of a 2-D image is obtained which can be enhanced etc. and displayed. The scan-time need only be 1 s or so. This type of equipment is widely used in airport security baggage search.

Group 3 equipment has not so far been used much for industrial NDT applications. In baggage search applications an array about 700 mm long containing 700 photodiodes is used, usually with the specimen scanning across a fixed array of detectors, by moving the baggage on a travelling belt. Each detector is therefore about 1 mm wide and can be made relatively thick so as to absorb more X-rays. Owing to the width and spacing of the detector elements, image definition is not good, and has usually been considered too poor for NDT flaw detection applications in welds and castings. Much finer linear arrays are now available, but are correspondingly thinner, so absorb a smaller percentage of the incident radiation and so produce noisier images. They may have industrial applications with low energy X-rays.

In addition to these three main groups of radioscopic equipment, there have been proposals for a TV camera with a target plate of selenium or lead oxide which is directly sensitive to X-rays. This could be operated in the same manner as a conventional photoconductive vidicon, but using X-rays instead of light as the incident radiation. Small tubes of this type with a pick-up surface 12×8 mm have been available for several years. With these tubes the final presentation of the image is on a large TV monitor screen so there is considerable magnification, and these direct X-ray sensitive tubes are capable of resolving very fine detail but require large incident dose-rates of X-rays of the order of 30–100 R min^{-1} to give the best results. Large-screen direct X-ray vidicons have been reported in the literature but are not yet available commercially [8].

There are various other possibilities, such as the use of microchannel plates, fibre-optic coupling, and the use of wide-aperture Schmidt mirrors instead of lenses, light amplifiers etc.

In general, gamma-ray sources are not much used for radioscopy because of their low radiation output, but equipment called Sentinel, marketed by Amersham International (UK), uses a collimated gadolinium-153 source spaced a few centimetres from a screen and a miniature TV camera as a hand-held real-time imaging system which is battery operated and weighs only 4.3 kg. The image is produced on a separate 100 mm monitor operated from a battery-pack carried on a belt. A resolution of 5 lines mm^{-1} is claimed.

11.4 IMAGE STORAGE AND IMAGE PROCESSING

With both Group 1 and 2 equipments, the image produced by the TV camera is usually converted to digital form, processed and modified in a computer before being presented on a TV monitor screen.

Most image processing systems today use a $512 \times 512 \times 8$ bit format, but $1024 \times 1024 \times 12$ bit is beginning to be used. It is important to match the image resolution of the X-ray image formed on the primary screen with the resolution of the image intensifier tube with the resolution of the TV camera and to the potential of the digital image processing system; most of these factors depend on the image size required. There is little point in using a high-definition image processing system, say 1024×1024 pixels, if the camera resolution is only 250 lines full-format.

Once the data are in digital form, they can be stored on tape or disc, as described in section 9.10. Increases in computer storage capacity now enable images to be stored on a very large pixel matrix, so that in many cases added image unsharpness due to pixel size is no greater than the original image unsharpness.

11.5 PERFORMANCE OF TELEVISION-FLUOROSCOPIC SYSTEMS (RADIOSCOPY)

The assessment of performance of any image intensifier system is complex, due to the number of image transfer stages involved. High resolution may be obtained at the expense of reduced performance on low-contrast detail, because an increase in sharpness will enhance noise. The problem of image quality in the output image is a problem of threshold detection: what is the minimum signal that can be detected in the presence of noise? Leaving aside instrumental problems such as distortion, instability, shading etc., the quality of a radiographic image can be assessed in terms of three parameters: sharpness, contrast and noise.

Image sharpness

All radiographic images are unsharp and the classic way of quantifying this is to measure the unsharpness width of the image of a physically sharp edge. Modulation transfer functions can also be used. In a television fluoroscopic image there are various causes of unsharpness which need to be measured and combined.

1. All X-ray-to-light conversion screens have a considerable unsharpness, typically at least 0.3 mm and sometimes much larger.
2. The fact that the image is formed by means of a pixel matrix is another cause of unsharpness, the size of which depends on how many pixels are used in relation to the length of the specimen that is imaged. For example, 512 pixels to 100 mm length of specimen means an unsharpness of 0.2 mm.
3. The image is presented on a TV line raster, standard European systems being 625 lines. If the image is real-size on the TV screen there

will be 625 lines for a 250 mm high specimen, so one line-pair will be 0.4 mm.

4. Television cameras have a limited resolving power which varies with the light input used and is usually quoted as the maximum number of TV lines, horizontal and vertical.

5. X-ray image intensifier tubes also have a maximum resolution which varies with the electronic magnification and also varies from centre to edge. It is usually quoted for 80 kV X-rays and may be less with higher energies: it is principally, but not wholly, dependent on the nature of the primary screen in the tube.

These causes of image unsharpness are inherent to the equipment and are in addition to geometric unsharpness due to the focal spot size of the X-ray tube.

Image unsharpnesses are not directly additive as was explained in Chapter 5, but the 'sum of the squares' rule gives an accurate value of the total effective unsharpness, and for good, well designed radioscopic equipment with no projective magnification, this is of the order of 0.5 mm, almost twice the unsharpness of the primary screen and several times larger than is recommended for good quality film radiography. As unsharpness is a major factor in radiographic flaw sensitivity, this is at first sight a serious limitation to the flaw detection capability of real-time radiography equipment.

The value of 0.5 mm represents the unsharpness due to the equipment, and because for thin specimens the geometric unsharpness is unlikely to be greater than 0.2 mm, it is usually the total effective unsharpness. One way to reduce this total unsharpness is to arrange for the image to be larger than natural size on the primary screen. If the image is M times natural size, the effective total unsharpness is reduced by M, provided of course that in raising M above one the geometric unsharpness is not also increased above the equipment unsharpness. One of the most powerful methods of doing this is to use a minifocus or microfocus X-ray tube with projective magnification. If an X-ray tube with a focal spot of 20 μm is used and the specimen is placed close to the tube with the primary screen at a distance, so that the image is four times natural size, the geometric unsharpness will still be only 0.06 mm, but the effective total unsharpness will be reduced to $0.5/4 = 0.13$ mm. The disadvantages of using projective magnification with a microfocus X-ray set are:

1. such sets are much more expensive;
2. They have a lower X-ray output as they must be run at lower currents (mA) and this may introduce other image quality problems;
3. the area of specimen examined at each view is reduced;
4. the detector-to-specimen distance needs to be equal to the (focal spot-to-specimen distance) M, which may be physically difficult.

For many applications the use of a minifocus X-ray tube with a smaller amount of projective magnification offers a useful compromise.

Image contrast

In film radiography there is usually a contrast enhancement factor, i.e. the film gradient, of about 4–6. Television cameras and image intensifier tubes do not increase image contrast, indeed the image conversion contrast is usually slightly less than one, but the use of image processing on a digital image makes the modification of contrast relatively easy. There are a number of contrast enhancement programs, both for overall enhancement and for local enhancement.

Image noise

The enormous amplification factor attainable through TV circuitry brings in a new factor which can limit performance. Briefly, the quality of the final image can be limited by statistical fluctuations in the number of X-ray quanta utilized in forming the image. The final image in any fluoroscopic system is a light image, consisting of a pattern of areas of differing brightness. If the brightness of an area of interest differs greatly from the surrounding area, the eye will recognize this area without difficulty. If the brightness difference is small, however, the perception of the area may be impossible, either because the brightness difference is below the minimum contrast threshold of the eye at the prevailing brightness, or because the brightness of the surrounding area is fluctuating. The origin of this background fluctuation is in the quantum nature of the radiation (both light and X-rays) and electrons.

The X-ray image on the primary screen is formed by the quanta transmitted through the specimen and absorbed in this screen. The screen usually absorbs only a small proportion of the X-ray quanta incident on it. Even if this primary image is formed from only a few quanta per square millimetre per second on the screen, by using intensifiers and TV circuitry it can still result in a bright final image, due to the amplification processes through the system, but at the same time the quantum noise in this primary image will also be amplified.

The laws of quantum fluctuations are fundamental and cannot be changed. If there are N quanta $mm^{-2} s^{-1}$ utilized in forming an image, the fluctuation (the noise) is $N^{1/2}$ and the signal-to-noise ratio is $(N^{1/2}/N)100 = N^{-1/2} \times 100$ (%). If, therefore, the image is produced from only a few X-ray quanta, the final image will be noisy and this image noise may obscure image detail. There are three ways of minimizing this limitation.

1. To use a primary conversion screen which will absorb and convert a greater proportion of the incident X-ray quanta.

2. To use a higher X-ray output or a higher-than-normal kilovoltage to get a higher intensity on the screen.
3. To integrate the signal over several seconds, instead of using only the number of quanta absorbed in the time of one TV frame. This can be done by a variety of methods in the image processing procedure and is one of the most powerful arguments for using digital image storage. If 25 frames are averaged, the time-delay need be only a little more than 1 s; if 100 frames are averaged, this is only about 4 s collection time, and the time-delay in producing the image will be less than 5 s. The signal-to-noise ratio can be shown to be proportional to $F^{1/2}$, where F is the number of frames used. Pedantially, the system is no longer 'real-time'.

Sturm and Morgan [9] proposed the following formula to describe the performance of a fluoroscopic system:

$$d = \frac{2K}{C} \left(\frac{g}{\pi t \, N_0} \right)^{1/2} \times 100 \qquad (11.1)$$

where d is the diameter of a circular image on the fluorescent screen (in millimetres), C is the contrast of this image, N_0 is the number of light quanta reaching the retina of the eye from each square millimetre of the fluoroscopic screen per second, t is the storage time of the eye and K is a constant between 3 and 5. Equation 11.1 then gives the smallest diameter image of contrast C in terms of the number of utilized quanta. If the image contrast is higher, a smaller diameter image can be detected, and vice-versa.

With an image intensifier system it can be shown that the useful amplification so far as detail sensitivity is concerned is that which brings the number of light quanta utilized on the retina equal to the number of X-ray quanta absorbed and utilized by the primary fluorescent screen. Many image intensifier systems have sufficient amplification to show quantum fluctuations; the image has a sort of 'boiling' appearance, and increasing the amplification does not improve the detail sensitivity.

11.6 IMAGE QUALITY MEASUREMENT

In any real-time radiography system the image is displayed on a TV monitor on a line raster. This image is of a completely different nature to that on a radiographic film, which usually has a grain structure barely visible without some magnification. It is not by any means certain, therefore, that conventional IQIs such as wires will give a fair or comparable representation of image quality. For example, in welds the most serious potential defect is the small crack, and it is not yet at all clear how well such defects will be imaged on a TV screen, even if shown at several times natural size. It would be valuable to know whether, when the wire IQI sensitivity is the same on a TV image as on film, the flaw sensitivities can also be equated for different types of flaws. Calculations of flaw sensitivity suggest that this is not so and that wire IQI sensitivity is not a satisfactory method of assessing the performance of a real-time radiography system. In particular, it appears possible to have an apparently good IQI sensitivity and at the same time a poor crack sensitivity.

Modulation transfer function methods

The method of assessing the performance of imaging systems on the basis of the concept of contrast transfer using spatial frequency as a parameter, as detailed in Chapter 8, is of great value in assessing X-ray image intensifier systems. Modulation transfer function (MTF) curves can be produced to represent each stage in the imaging system, primary screen, optics, TV camera etc., and are easily combined to produce a resultant curve for the complete system (Fig. 11.5). The usefulness of these curves is that they give a clear indication of the limiting components of a system. Figure 11.5 shows a family of MTF curves for the components of one type of radioscopic system. It is clear from these data that the primary conver-

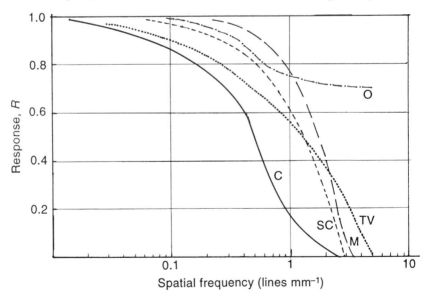

Fig. 11.5 Modulation transfer function curves for a radioscopic system and its components. Curve 0 = optics; curve TV = television camera tube; curve M = display monitor; curve SC = conversion screen; curve C = combined curve representing the performance of the complete system.

sion screen (X-rays to light) is the limiting parameter and that the optical system has a much higher performance than is necessary.

Most MTF data are normalized to 100% response at zero frequency, and this may disguise the impact of some system parameters if the contrast transfer across a particular stage of the process is not unity at zero frequency. In spite of such limitations, however, the cascading of MTF values for the individual parameters of a system does yield valuable insight into system performance.

There have been attempts to convert MTF data into a single value 'quality criterion'. Schade [10] introduced the concept of a 'noise equivalent pass-band' (N_e), defined as

$$N_e = \int (MTF)^2 \, df \tag{11.2}$$

and showed that this correlates well with the general subjective impression of image sharpness. He proposed that equation 11.2 be used as a criterion of the performance of TV image systems.

11.7 PRACTICAL PROCEDURES

Depending on the image quality and flaw sensitivity requirements of a specific application, a decision must be made on the acceptable total unsharpness, referred back to the size of the specimen. If the specimens are welds in which small flaws such as cracks might occur, it is likely that an unsharpness comparable with that attained in film radiography will be needed.

It is recommended that in applications where a high sensitivity is required for the detection of planar flaws such as cracks, lack of fusion or narrow lack of penetration in welds, that a criterion of satisfactory image quality should be more than a conventional wire or hole IQI sensitivity value [11,12]. An additional measure of image sharpness is required. This can conveniently be provided by the image of a duplex wire IQI such as type IIIA in BS:3971:1985 (EN–462–5) (Fig. 7.3). Theoretical studies of radiographic sensitivity show that wire IQI is strongly dependent on contrast parameters, whereas crack sensitivity is much more dependent on the image unsharpness which, as has been shown, is comparatively large with present-day radioscopic systems.

For the inspection of castings and assemblies, when the likely flaws are larger, a larger image unsharpness may be acceptable.

These considerations will decide whether a conventional X-ray set with a focal spot in the range 2–4 mm diameter used with no projective magnification can be used, or whether a microfocus or minifocus tube with projective magnification is necessary.

11.8 IMAGE PROCESSING

It is unlikely that the unprocessed single-frame image will be viewed directly on the TV monitor screen. With a $512 \times 512 \times 8$ bit pixel matrix (12 bit is to be preferred for signal-to-noise ratio improvement) modern digital image processing equipment can provide a standard image enhancement program which takes only a few seconds. A typical program would:

1. reduce quantum noise by frame integration, over (say) 32 frames;
2. provide thickness equalization across the image;
3. provide contrast enhancement (this can be either local or overall, as required);
4. Provide image 'crispening' (a form of improvement in apparent image sharpness by edge enhancement).

In addition, there is a range of image enhancement programs such as histogram equalization, difference imaging, displacement differencing etc. to deal with difficult images.

So much IQI data are available for film radiography that for industrial applications it is natural to require comparison data for radioscopic systems, but there is such a range of equipment with refinements such as enlarged images and image processing that no overall comparison data are easily possible. IQI sensitivities as good as normal film radiography are attainable on many specimen thicknesses. The limit to performance is usually the primary conversion screen: for thin specimens it is an unsharpness problem and for thicker specimens a quantum fluctuation problem.

11.9 APPLICATIONS

Kobayashi [13] reported the use of an image intensifier in a shipyard for routine inspection of submerged-arc welds in 14 mm steel plates, with an inspection speed of $1m \ min^{-1}$. More recently, the potential advantages of a digital image have led to a rapid increase in the reported uses of radioscopy.

Radioscopic systems have been used for preliminary sorting of castings such as automobile wheels, using a remote-handling specimen manipulator, and proposals have been made that such a system could incorporate computerized image interpretation and become a completely automatic inspection system [14, 15].

Radioscopic systems have also been used for the inspection of large rubber tyres to check the correctness of metal inserts. Similar systems, with a manipulator, are being used for high-speed weld inspection on pipe-welds on lay barges and in construction yards [16]. Very large equipments using linacs or large gamma-ray sources (600 Ci cobalt-60) have been used to inspect rocket motors [17] and drums of waste radioactive material [17], while microfocus X-rays and a magnified image with a 1024×1024 pixel matrix has been used to monitor welded repairs on turbine blade tips [18].

There have been some interesting applications to examine moving systems. Either the image can be viewed on the screen while the event is happening, or it can be filmed or recorded on video tape; if recorded, it can be used for slow playback or the study of single frames. The pouring of liquid metal into a casting mould to study the flow of metal through the ingate, the filling of the mould cavity and the formation of casting flaws during solidification have all been studied in steel, aluminium, and some non-metallic materials [19–22]. Under ideal conditions, when maximum radiographic sensitivity is attained, it may be possible to discern the liquid–solid interface [21]. The flow of liquids, the presence of cavitation etc., has also been studied. Greater filming speeds are possible with simple image intensifier tubes than with a TV-coupled system, by using a high-speed camera connected directly to the intensifier output phosphor.

Another interesting application involved the electroslag welding process [23] to study the events within the slag bath; the effects of misalignment of the consumable wire were clearly demonstrated.

At Rolls-Royce, Stewart [24] pioneered the use of an 8 MeV linear accelerator to radiograph a full-size aero-engine while this was running on a test-bed, to study the movement of seals. In 1972 he extended this work [25] to use high-energy television-fluoroscopy for the same purpose. Since then, other aero-engine manufacturers have developed similar techniques. The most useful feature of this application is the ability to determine the points in time when certain mechanical events occur during engine operation, and to study the rate-of-change of component positions within the engine. Some related work was recently reported using an image intensifier tube with a primary screen sensitive to neutrons [26] to study oil flow in an engine.

Securely baggage search is a major application of radioscopic equipment (section 11.3). For normal airport luggage a medium-energy X-ray set is adequate, but much larger equipments have also been built to deal with large containers. The main problem is the interpretation of images of complex items in luggage. A short pulse of X-rays is used to reduce the X-ray dose to a photographic, film-safe level. In some equipment, two X-ray beams mutually at right-angles are used to obtain all dimensions of a suspect item. Very large equipments using the same basic principles (Group 1) have been built, using one or two 8 MeV linacs and a very large screen, to examine container lorries at international frontiers. The vehicle is traversed slowly across a fluoroscopic screen 250 cm wide and the image is taken on to several TV cameras simultaneously. Usually such radioscopic installations have been combined with sophisticated explosive/alcohol/drugs detectors (sniffers).

11.10 AUTOMATED IMAGE INTERPRETATION

Once the image is held in a computer store, pattern recognition techniques can be used to accept or reject images automatically according to pre-set criteria. The human inspector need not look at the images except in borderline cases. Computer programs are being developed for this purpose in a number of inspection fields [27–29]. Either the flaws can be described by type/location/size/orientation, or if there is an agreed defect acceptance criteria, a simple accept/reject decision can be made by the computer. Alternatively, a library of rejectable defects can be held by the computer, for comparison. Automated image interpretation is a complex subject and a universal system is not possible. One potential method [29] is to enhance the image, eliminate irrelevant indications, and then measure each image in terms of a number of parameters: flatness, symmetry, unsharpness, non-linearity, length, width, angle and distance from the weld centre. An inspector's 'knowledge-base' is constructed in the computer and this is the most difficult task, as many inspectors solve flaw interpretation problems intuitively, based on information other than is shown on the radiograph.

A considerable literature is now available on automated image interpretation as the subject is not limited to radiography or to weld inspection [30, 31].

11.11 STANDARDIZATION IN RADIOSCOPY

ASTM has published a series of guidance documents on radioscopic methods [32–34] which take the form of extensive monographs on equipment, procedures, recommendations on standardization and weld inspection. A standard on 'The application of real-time radiography to weld inspection' (BS:7009:1988) [12] covered methods of testing equipment with a test-card, and performance standards in radiographic terms. Draft standards have also been prepared by CEN and ISO [35].

11.12 FUTURE DEVELOPMENTS

Apart from weld inspection, much of the interest is in the application of radioscopic systems outside the conventional NDT field.

The inspection of baggage for airport security has already been mentioned, with pulsed X-rays and image storage to reduce the radiation dose to the package. A completely different system of baggage inspection with X-rays, which has also been used for field medical radiography, is usually known as the 'flying spot' method [36]. A radial slit in a disc is rotated in front of a fixed vertical slit so that a small spot of X-rays is scanned rapidly along a vertical line. If a specimen (the suitcase etc.) is then moved horizontally across this scanning spot, the effect is to scan a small disc of X-rays over the specimen, in the form of a raster. The transmitted X-rays are detected on a strip of sodium iodide scintillator which is the length of the vertical scan and which is coupled to one or more photomultiplier tubes. The signal is exactly analogous to a TV signal from a camera tube, but with a relatively slow scan-time, and can be converted to an image on a display monitor. For baggage inspection the particular merit of the system is that because the signal is detected in a thick crystal, all the transmitted X-ray energy is utilized, leading to a very efficient conversion process. As the X-ray beam is scanned rapidly, the radiation dose to any part of the specimen is very small (typically $50\,\mu\text{rad}$), and the inspection does not damage photographic material which may be in the luggage.

For other applications the size of the scanning spot could be reduced, and if the horizontal traverse speed of the specimen was also slowed, there would be more lines per image width, i.e. better definition. So far as is known, this possibility has not yet been investigated. The long strip of single-crystal scintillator could be replaced by a linear matrix of small crystals or semiconductors and this would enable the flying spot of X-rays to be replaced by a stationary slit beam.

Another method of X-ray imaging, which is a hybrid of TV and a large imaging surface, is the Digiray equipment developed by Albert [37–40]. In this, a special form of TV tube is used in which the electron beam is scanned over a large-area target, producing a small ($25\,\mu\text{m}$) moving spot of X-rays. The specimen is placed close to the tube with the detector at a distance, so that scatter in the specimen is minimized and a fan beam of X-rays is produced. By synchronizing the generator and display rasters, an image is built up digitally in a computer memory. As the detector can be a thick crystal (NaI), it can be almost 100% efficient at low X-ray energies, and the image resolution depends on the X-ray beam diameter. Due to the absence of scattered radiation at the detector, contrast sensitivities as good as 0.2% are claimed. At present the Digiray tubes operate up to 100 kV, and aluminium specimens up to 65 mm thick and steel up to 13 mm have been examined. By using two detectors, stereo-imaging is claimed to be possible.

11.13 COMPUTED TOMOGRAPHY

A whole new family of imaging processes has been developed from the concept of computer processing an array of stored X-ray absorption data.

These are the well-known 'brain scan' and 'body scan' tomographic equipments invented by Hounsfield [41]. Nearly all the basic work was done in the medical field, but some of the techniques are now being used for industrial applications. In the medical field there is an extensive literature on equipment and uses, but for industrial purposes the requirements are somewhat different. A comprehensive literature survey of industrial developments is available [42].

The basic principle of computed tomography (CT), also known as CAT, is as follows (Fig. 11.6). If a small pencil of X-rays is passed through a specimen, the transmitted radiation can be detected and measured on the output side of the specimen with a small detector such as a scintillation crystal or a semiconductor, and this measurement is then digitized and stored. In practice a linear array of several detectors is used. If the specimen is rotated in steps and also moved transversely between the X-ray source and detector until the whole specimen volume is scanned, with a reading taken at each position, a large quantity of attenuation data is obtained, each element in the specimen having contributed to several attenuation readings. The specimen movement must be extremely accurate so that each element is accurately located in space, particularly along the translational axis.

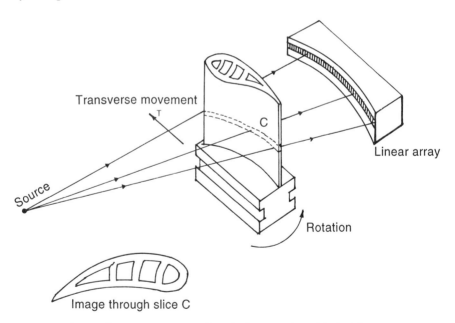

Fig. 11.6 Set-up for X-ray tomography of a large turbine blade. The specimen is rotated and traversed across a fan of X-rays from a source. The transmitted radiation is detected on a linear array. Below is the image of slice C through the blade.

After complete rotation and traverse, the data can be processed in a computer to build an image of a thin slice of the specimen, cut along the plane of the detector array, of thickness equal to the thickness of the detectors. The image obtained is as if a thin slice has been cut from the specimen, this slice laid on a film, and a radiograph taken of the slice (Fig. 11.6). The image is therefore quite different to a conventional radiograph through the specimen. One thin slice of the specimen has to be imaged at a time. In medical tomography the X-ray source and the detector assembly are rotated around a stationary patient and the emphasis of the equipment design is on keeping the radiation dose to the patient to a minimum. In industrial applications the radiation dose to the specimen is unimportant; also, it is usually more convenient to move the specimen between a fixed X-ray source and the detector array.

To reconstruct a 2-D map of X-ray linear attenuation coefficients (i.e. an image of the slice), two types of algorithm have been used. In one approach an image is obtained by back-projecting transmission data convolved with a spatial frequency filter; equivalently, the same recon-struction can be obtained by a 2-D Fourier transform of the function of the projections arranged along appropriate diagonals in frequency space [43]. The second approach begins with an arbitrarily chosen sectional distribution which is then adjusted iteratively until the projections of the adjusted picture match the measured projections, subject to a criterion of acceptability. There is an enormous literature describing these algo-rithms in the medical field of CT [44–46] and each method has special advantages in particular applications. To reduce image reconstruction times to practical values, an array processor or a dedicated hard-wired system is necessary, even though in industrial applications the specimen is normally static.

More recently, new algorithms have been proposed [45] to reduce image reconstruction time. Proposals have also been made to reconstruct images from a limited number of scan positions using 'sparse data algorithms'. These, of course, result in poorer quality images, but may be good enough for specific applications. New detector materials have been proposed for use with high energy X-rays, e.g. cadmium telluride [47].

All CT systems are limited in their performance by many of the par-ameters affecting conventional radiographic imaging systems: finite source size, source-to-specimen distance, specimen-to-detector distance, detector diameter (or gate) [48]. In practice, the specimen-to-detector distance may be considerable and the image is almost all penumbra, with no umbra (as in Fig. 6.5(c)). By using MTF methods, as described in Chapter 8, the limitations can be built into an assessment of the overall performance of the equipment. The second major limitation to performance is noise: statistical noise due to quantum fluctuations; electronic noise; noise from the digitization process, the reconstruction process; and detector/time limitations.

There are additional problems in CT with pulsed X-ray sources such as linacs, in that the individual spot measurements from specimen displacement positions must be synchronized with the emission of the X-ray pulses. The pulsed emission, however, allows one to make baseline corrections between two measurements, to overcome variations in X-ray output intensity.

Two other developments in equipment are the following:

1. a cone beam of X-rays and a 2-D array of detectors eliminates misalignment problems and gives faster data acquisition [49];
2. a dual-energy beam utilizes the difference in attenuation to identify specific materials in a complex specimen.

After a period when the industrial acceptance and use of CT equipment seemed to be slow, there have been many new reports of successful applications using the whole range of radiographic sources from microfocus X-rays through to betatrons and linacs [50–58]. The problems of using pulsed X-ray sources have been solved and tomographic equipment using large linacs (8–15 MeV) is in use for the inspection of solid rocket motors and radioactive waste containers [17, 58].

As computers are able to store more data and there are further improvements in software, the cost and time to produce a tomographic image are reducing: 512×512 pixel images are now common, and the use of 1024×1024 pixels has been reported [59]. All the usual techniques of digital image enhancement, image storage etc., as applied to radioscopic images, are possible. The quality of the resultant image depends on the source/detector geometry and the total number of sampled readings which comprise angular steps around the specimen, together with the linear diametric steps. The size of the specimen determines how many are needed for a given resolution, since the source/detector effective beam width must sample sufficient elements at the circumference if the image quality is to be maintained. For small objects and 0.5 mm resolution, fewer than 10^6 readings are sufficient. The image reconstruction time is dependent on the capacity of the computer, and for fast times a special processor is necessary. Typically, modern equipment might require 5 min scan time and 10 min reconstruction time, but this obviously depends on specimen size, pixel matrix and computing capacity.

Microtomography of very small specimens is possible by using a microfocus X-ray tube and projective magnification. Mobile CT equipment, using a gamma-ray source, has been built for site use [51]. The limited flux from a gamma-ray source means that image quality requires a longer scan time, and the benefits of portability have to be offset against the extended scanning time and poorer resolution.

Timber logs have been inspected for metal inserts with modified medical CT equipment. It has been reported that by using a coarse pixel raster and a very limited number of scan positions, one image per second can be captured and reconstructed, with sufficient image detail to satisfy requirements [60].

An interesting new development in CT is that in conventional radioscopic television equipment, where the whole area of a specimen is imaged, this image can be obtained and stored at a series of positions as the specimen is rotated. These stored images can then be processed to produce tomographic images of any chosen slice of the specimen. A very large computer store is obviously necessary and the number of steps of rotation is unlikely to be as large as would be used in a conventional CT equipment, nevertheless reasonable quality images have been obtained. Results have been published [61] which used a 180° rotation in 2° steps. The important advantage of this technique is that only one series of exposures is necessary to produce data for a large number of tomographic slices.

Another feature of CT is that the attenuation data obtained enable the physical density of any part of the specimen to be obtained with high accuracy (0.02%).

There are two ASTM standards in the form of guides for X-ray tomography [62, 63].

Computerized tomography is not limited to X-rays and gamma-rays. The use of radioactive liquids, positrons and neutrons has also been demonstrated [64], using appropriate detectors. Neutron tomography has applications to fibre/composite structures, explosive-filled stores and nuclear fuel elements. Positron tomography can be used to detect liquids inside metal structures.

11.14 BACKSCATTER METHODS

In recent years several methods have been proposed for using the intensity of backscattered radiation to produce tomographic-type images. The basic principle is shown in Fig. 11.7, in which an array of detectors is scanned across a specimen and a suitable computer program is used to construct a tomographic image. The most promising applications are to low-density specimens, using radiation where Compton scattering is predominant. A 'slice' image is obtained with the plane of the slice at right-angles to the radiation beam, that is, parallel to the specimen surface.

Holt and Cooper [65] attempted a comparison of computer simulation of Compton scatter at 90° and experimental data, using monoenergetic

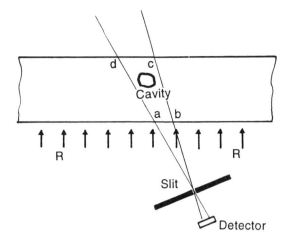

Fig. 11.7 Principle of the Compton back-scatter technique. The detector receives scatter from volume abcd of the specimen, through the slit. If there is a cavity in this volume the intensity of the backscatter changes as the specimen moves across the radiation field R–R.

gamma-rays from Cs-137 (energy 0.662 MeV). They showed that the image contrast produced by a void in the specimen is effectively independent of the incident radiation energy, that a single scattering approximation is adequate to explain the experimental results, and that a void will be detectable if each of its linear dimensions is greater than half the beam width, that is, the void is greater than about 10% of the scattering volume V. This implies that to detect smaller voids the scattering volume must be reduced, and this increases the scanning time by a factor of V^2. The work was done on a range of materials, including light-alloys and plastics.

A theory was proposed [66] which shows that the relative sensitivities of transmission tomography S_t and scatter tomography S_s are given by

$$S_t/S_s = (1/\mu L)^{1/2}$$

where L is the specimen thickness and μ is the linear attenuation coefficient. Thus, for thin, low-density specimens, as $\mu L < 1$, backscatter tomography may be more sensitive for flaw detection. A practical advantage of backscatter methods is that both radiation source and detector can be on the same side of the specimen and access to the opposite side is not necessary.

As the radiation detector is a crystal, its thickness can be matched to the radiation energy employed, and the radiation detection efficiency can be higher than a film emulsion. There is a complex relationship between

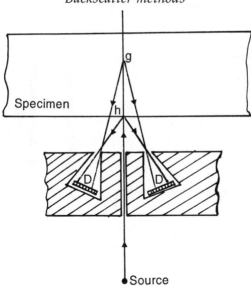

Fig. 11.8 One element of the detector array proposed by Kosanetsky [66]. Individual elements in each array D receive scattered radiation from layers at different depths in the specimen.

source strength, detector size, equipment geometry, the statistics of counting, the inspection speed and the volume of the minimum detectable flaw in a specimen [67]. Present theories suggest a square-law dependence between inspection speed and minimum detectable void volume. Compton theory is complex and multiple scatter problems require the use of Monte Carlo methods for rigorous solutions.

Practical, published results [68] suggest that very high-strength sources are necessary, for fast inspection speeds on large specimens, but some published laboratory results have used a low-strength source of Ba-133 (1.4 mC) on ceramic specimens [69].

The advantages of backscatter tomographic methods are that they can be applied to very thick specimens where transmission methods would be impossible, although it would seem at present that useful images may be limited to near-surface regions. Three-dimensional depth information is possible by using an array of oriented collimators, and access to one side only is required (Fig. 11.8).

Forward-scatter techniques simplify the equipment-shielding problem and might have applications to smaller specimens and automated systems. Continuous monitoring applications for the detection of corrosion or erosion etc. do not necessarily require the imaging of small detail, and for such applications a small source and relatively portable equipment may be practical [70]. By using a slot camera (Fig. 11.9) a system with a

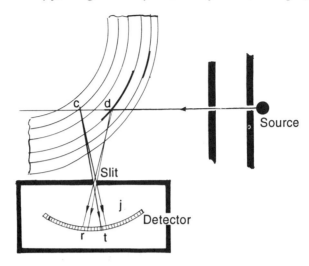

Fig. 11.9 Compton backscatter system [71]. Different elements on the detector collect radiation from different layers (c, d) in the specimen through a slit.

much faster response has been claimed. The slot effectively images different layers in the specimen at different points on the film. If there is less scattered radiation from a small volume at a particular point d, there will be a reduced intensity at r′ on the detector plane. Initially, film and intensifying screens were used as a detector [71, 72] and the image was examined with a microdensitometer, but these have been replaced by an array of rectangular-shaped detectors providing digital data. The film/screen system required exposure times of 1–2 h. It has been demonstrated that a slot camera with a film/screen detector can show the presence of artificial laminations of widths down to 0.15 mm or less in Perspex specimens, and 'images' of layer separations at more than 70 mm below the surface have been detected. The slot camera is potentially much faster than a pinhole camera approach and a scintillation detector should be much faster than film. The detection of very narrow laminations by backscatter methods, when the volume change over a finite volume of specimen is relatively small, is unexpected: a 0.1 mm separation inside a 2 mm cube of material represents only a 5% difference in volume. Again, one major advantage of the slot camera system is that access to only one side of the specimen is necessary, so that very thick specimens such as solid rocket motors can be examined.

Pipelines can be examined through outer layers of insulation to detect metal thinning. Using 160 kV X-rays, steel thickness changes of 5% have been detected in an 8 mm steel wall [50]. A 90° scatter method has been reported for density measurements, claiming an accuracy of 0.5% in the density determination [73]. A further development of this backscatter technique is to use a flying spot of X-rays [74] and several radiation detectors. This enables an area to be scanned without moving the specimen and permits the use of a large-area detector.

11.15 LAMINOGRAPHY

It is possible to perform the equivalent of classical tomography by digital means, instead of using film. A similar but simpler apparatus to CT can be used, in which the source and detector are both moved around a specimen. The result is a digital radiograph showing an area image but with overlying planes removed. The technique has some of the benefits of CT and some of conventional radiography [75, 76]. Alternatively, the film and specimen can be rotated synchronously with a stationary source, and the blurred image deconvolved through a regularizing Fourier filtration [77].

REFERENCES

[1] Halmshaw, R. and Hunt, C. A. (1957) *Brit. J. Appl. Phys.*, **8**(7), 282.
[2] Teves, M. C. and Tol, T. (1952) *Philips Tech. Rev.*, **14**, 33.
[3] Coltman, J. W. (1948) *Radiology*, **51**, 539.
[4] Placious, R. C. and Polansky, D. (1991) *Mater. Eval.*, **49**(11), 1419.
[5] Burakin, V. N. (1974) *Defektoscopia*, **10**(3), 142.
[6] Berger, H. (1994) *Mater. Eval.*, **52**(11), 1248.
[7] Astromed, Cambridge, UK (1994) Data sheets.
[8] Nudelman, S. and Ouimette, D. R. (1993) *US Patent 5195188*.
[9] Sturm, R. E. and Morgan, R. H. (1949) *Am. J. Roentgenol.*, **62**, 617.
[10] Schade, O. (1964) *J. Soc. Motion Pict. Tel. Eng.*, **72**(2), 81.
[11] International Institute of Welding (1990) *Brit. J. NDT*, **32**(1), 9.
[12] British Standards Institution, (1988) *Application of Real-time Radiography to Weld Inspection*, BS 7009.
[13] Kobayashi, K. (1968) *Brit. J. NDT*, **10**(4), 70.
[14] Purschke, M. and Schaefer, M. (1992) *Proc. 13th World Conf. NDT*, Sao Paulo, Brazil. Elsevier, Vol **1**, p. 572.
[15] Purschke, M. and Schulenburg, H. (1994) *Proc. 6th Euro. Conf. NDT*, Nice, France. COFREND, Paris, Vol. **2**, p. 1249.
[16] Oilfield Inspection Services, Great Yarmouth, UK (1992) Technical literature.
[17] Lamarque, P. (1994) *Proc. 6th Euro. Conf. NDT*, Nice, France. COFREND, Paris, Vol. **2**, p. 1251.
[18] Nuding, W., Link, R. *et al. Proc. 6th Euro. Conf. NDT*, Nice, France. COFREND, Paris, Vol. **1**, p. 683.
[19] Halmshaw, R. and Lavender, J. D. (1966) *Iron & Steel, London*, **39**(9), 381.
[20] Hall, H. T. (1967) *Metals*, **2**(18), 50.
[21] Forsten, J. and Mielkka-Oja, H. M. (1967) *J. Inst. Metals*, **95**, 143.
[22] Stegemann, D., Reimche, W. and Schmidbauer, J. (1992) *Euro. J. NDT*, **1**(3), 107.
[23] Lowery, J. F. (1968) *BWRA Rep. P/32/68*.
[24] Stewart, P. A. E. (1972) *Chart. Mech. Eng.*, **19**(5), 65.
[25] Stewart, P. A. E. (1975) *Aeronaut. J.*, **79**, 331.

[26] Stewart, P. A. E. (1980) *ASTM Spec. Tech. Publ. No. 716*

[27] Daum, W. and Heidt, H. (1989) in *X-ray Real-time Radiography and Image Processing* (ed. R. Halmshaw), British Institute of NDT Northampton, Chapter 9.

[28] Daum, W., Rose, P., Heidt, H. and Builtjes, J. H. (1987) *Proc. 4th Euro. Conf. NDT*, London. Pergamon Press.

[29] Kato, Y. *et al.* (1992) *Welding in the World*, **30**(7/8), 182.

[30] Olsen, R. (1988) *Mater. Eval.*, **46**, 1403.

[31] Inoue, K. and Kobayashi, M. (1982) *Trans. Jap. Weld. Res. Inst.*, **11**(2).

[32] American Society for Testing and Materials (1994) *E–1000–92*. Standard Guide for Radioscopy, in *ASTM Annual Book of Standards*.

[33] American Society for Testing and Materials (1991) *E–1411–91 Qualification of Radioscopic Systems*.

[34] American Society for Testing and Materials (1992) *E–1416–92 Radioscopic Examination of Weldments*.

[35] International Institute of Welding (1991) *VA–449–61, VA–450–91 (1991) Qualification of Radioscopic Systems, Parts 1.2.*

[36] Stein, J. A. and Swift, R. D. (1972) *Mater. Eval.*, **30**(7), 137.

[37] Albert, R. D. (1989) *Proc. Real-time Topical III Conf.*, Cincinatti, OH, USA. American Society For Non-destructive Testing.

[38] Albert, T. H. (1993) *Mater. Eval.*, **51**(9), 1020.

[39] Albert, T. H. and Albert, R. D. (1993) *Mater. Eval.*, **51**(10), 1350.

[40] Birt, E. A., Parker, F. R. and Winfree, W. P. (1994) *Review of Progress in Quantitative NDE* (eds D. O. Thompson, and D. E. Chimenti), Plenum Press Vol. 13B, p. 1963.

[41] Hounsfield, G. N. (1979) *Phil. Trans. Roy. Soc. Lond.*, **292A**, 223.

[42] Anon. (1994) *NDT&E Int.*, **27**(2), 101.

[43] Webb, S. (1981) in *Physical Aspects of Medical Imaging* (eds B. M. Moores, R. P. Parker and B. R. Pullan), Wiley, Chapter 6.

[44] Gordon, R. (1974) *Int. Rev. Cytology*, **38**, 111.

[45] Natterer, F. (1986) *The Mathematics of Computerised Tomography* Wiley, New York.

[46] Munshi, P., Rathmore, R. and Vijay, S. (1994) *Euro. J. NDT*, **3**(3), 104.

[47] Glasser, F., Girard, J. L. and Lamarque, P. (1994) *Proc. 6th Euro. Conf. NDT*, Nice, France. COFREND, Paris, Vol. 1, p. 645.

[48] Burstein, P. (1990) *Mater. Eval.*, **48**(5), 579.

[49] Jan, M. L. (1994) *NDT&E Int.*, **27**(2), 83.

[50] Hunt, C. A. (1987) *Proc. 4th Euro. Conf. NDT*, London. Pergamon Press. Vol. **3**, p. 2105.

[51] Link, R. and Nuding, W. (1987) *Proc. 4th Euro. Conf. NDT*, London. Pergamon Press, Vol. 3, p. 2142.

[52] Drake, S. G. (1993) *Brit. J. NDT*, **35**(10), 580.

[53] Goebbels, J. and Heidt, H. (1987) *4th Euro. Conf. NDT*, London. Pergamon Press, Vol. 3, p. 2111.

[54] Barbier, P. (1994) *Proc. 6th Euro. Conf. NDT*, Nice, France. COFREND, Paris, Vol. 1, p. 651.

[55] Filinov, V. N. and Kluyev, V. V. *et al.* (1994) *Proc. 6th Euro. Conf. NDT*, Nice, France. COFREND, Paris, Vol. 2, p. 1243.

[56] Baranov, V., Kroening, M. and Morgner, W. (1994) *Proc. 6th Euro. Conf. NDT*, Nice, France. COFREND, Paris, Vol. 2, p. 1287.

[57] Martin, M., Sire, P. and Rizo, P. (1994) *Proc. 6th Euro. Conf. NDT*, Nice, France. COFREND, Paris, Vol. 2, p. 1293.

[58] Lierse, Ch., Gobel, H. *et al.* (1994) *Proc. 6th Euro. Conf. NDT*, Nice, France. COFREND, Paris, Vol. **2**, p. 1299.

[59] Ridyard, J. N. A. and Moore, J. F. (1988) *Proc. Portsmouth Conf. NDT*, British Institute of NDT, Plenum Press, p. 464.

[60] Schmoldt, D. L., Zhu, D. and Conners, R. W. (1993) *Review of Progress in Quantitative NDE* (eds D. O. Thompson and D. E. Chimenti), Vol. **12B**, p. 2257.

[61] Burch, S. F. (1988) *Proc. Conf. Real-time Radiography & Image Processing*, Newbury. British Institute of NDT.

[62] American Society for Testing and Materials (1993) *E–1441–93. Standard Guide to Computed Tomography Imaging.*

[63] American Society for Testing and Materials (1993) *E–1570–93. Standard Practice for Computed Tomography Examination.*

[64] Glasser, F. (1988) *Symp. Industrial Tomography*, Grenoble. France.

[65] Holt, R. S. and Cooper, M. J. (1988) *Brit. J. NDT*, **30**(2), 75.

[66] Kosanetsky, J. and Harding, G. (1987) *Proc. 4th Euro. Conf. NDT*, London. Pergamon Press, Vol. **3**, p. 2118.

[67] Bridge, B. (1985) *Brit. J. NDT*, **27**(6), 357.

[68] Bridge, B. (1986) *Brit. J. NDT*, **28**(4), 216.

[69] Bridge, B. and Bin-Saffiey, H. (1988) *Brit. J. NDT*, **30**(6), 392.

[70] Ong, P. S., Anderson, W. L. *et al.* (1993) *Review of Progress in Quantitative NDE* (eds D. O. Thompson and D. E. Chimenti), Plenum Press, Vol. 12A, p. 295.

[71] Berger, H., Cheng, Y. T. and Criscuolo, E. L. (1985) *NSWC Rep. TR-65.*

[72] Berger, H., Cheng, Y. T. and Criscuolo, E. L. (1987) *5th Pan-Pacific Conf. NDT* Vancouver, Canada.

[73] Muller, J., Ackermann *et al.* (1994) *Proc. 6th Euro. Conf. NDT*, Nice, France. COFREND, Paris, Vol. 2, p. 1201.

[74] Bossi, R., Friddell, K. D. and Nelson, J. M. (1988) *Mater. Eval.*, **46**(10), 1462.

[75] Danovitch, Z. and Segal, Y. (1994) *NDT&E Int.*, **27**(3), 123.

[76] Icord, C., Rizo, P. and Dinten, J.-M. (1994) *Proc. 6th Euro. Conf. NDT*, Nice, France. COFREND, Paris, Vol. 1, p. 639.

[77] Buchele, S. F., Ellinger, H. and Hopkins, F. (1990) *Mater. Eval.*, **48**(5), 618.

12

Special methods

12.1 INTRODUCTION

There are a few special methods or applications of radiology which have not yet been covered. The most important is the use of atomic particles, such as electrons, neutrons and protons, to form 'radiographic' images. Probably the most important of these is the use of neutrons.

12.2 NEUTRON RADIOGRAPHY

Neutrons are uncharged atomic particles of similar mass to the proton, which will penetrate matter and are attenuated in their passage, and so can be used to produce radiographs. It is usual to classify neutrons into four groups:

- fast neutrons, with energies exceeding 0.1 MeV;
- epithermal neutrons, with an energy range of $0.3-10^2$ eV;
- thermal neutrons, with an energy range of 0.01–0.3 eV;
- cold neutrons, with an energy range of 0–0.01 eV.

Thermal neutrons are so-called because they are in thermal equilibrium with their surroundings at, or near, room temperature. The properties of neutrons depend on their energy, and practically all neutron radiography, to date, has been done with thermal neutrons because of their interesting absorption properties.

Whereas the attenuation of X-rays in materials increases with increasing atomic number of the absorbing material, the mass attenuation coefficients of the elements for thermal neutrons, if arranged in order of increasing atomic number of the absorber, appear almost completely random. Neutron absorption does not depend on the electron structure of the atom as does absorption with X-rays, but on interaction with the atomic nucleus, and certain light elements such as hydrogen, lithium

Table 12.1 Mass attenuation coefficients for thermal neutrons

Element	Attenuation coefficient (cm^2 g^{-1})
Hydrogen	48.5
Boron-10	121
Aluminium	0.036
Iron	0.141
Cadmium	11.2
Gadolinium	84.0
Lead	0.034
Uranium	0.033

and boron, and some rare-earth elements such as gadolinium, dysprosium and indium, have high or very high thermal neutron absorption, whereas some of the heavy elements such as iron and lead have low neutron absorption. Neutron absorption coefficients may also vary between different isotopes of the same element. Table 12.1 shows typical values for some elements. These values indicate the potential advantages of radiography with thermal neutrons. By far the most important case is that of hydrogen, where the absorption is so high that there is the strong possibility of imaging hydrogenous components (plastics, water, explosives, oil etc.) inside metal structures. One of the earliest studies of neutron radiography imaged a piece of waxed string laid across a 5 cm thick lead block [1].

With fast neutrons the absorption coefficients of different elements are less separated and the prospects for fast neutron radiography seem much more limited, but epithermal neutrons offer some possibilities which do not yet seem to have been much studied, and cold neutrons appear to be capable of even better discrimination than thermal neutrons because they produce less scatter. In general, neutrons of higher energy than thermal neutrons will give greater penetration but with more scattering, so producing poorer images. To date, almost all industrial neutron radiography has used thermal or cold neutrons.

Neutron recording methods

Although standard photographic film will detect neutrons, the use of various types of converter screen results in a much more efficient recording process. The methods used are described below.

1. Direct-exposure metal foils. These are foils which, placed in contact with the film during the actual exposure, emit radiation or electrons while being bombarded with neutrons. Materials used are gadolinium, cadmium and lithium. For the best results a single 0.025 mm back screen of gadolinium should be used.

2. Direct-exposure granular screens. These are screens which convert neutron energy to visible light. The usual material is lithium-6 mixed with zinc sulphide. The granular screens are used with light-sensitive films, and are the fastest type of converter screen, permitting the shortest exposure times. The images obtained are, however, very grainy and therefore have relatively poor image quality. An alternative direct-exposure screen is a layer of gadolinium oxysulphide in a binder. The image is less grainy, but such screens are sensitive to any gamma-rays accompanying the neutrons, so that a 'mixed' (neutron and gamma-ray) image is obtained.

3. Direct-exposure scintillator glass. This is a glass-like material, usually cerium-activated silicate glass containing lithium enriched with lithium-6. It produces a visible light image which is less grainy than granular scintillator screens. Glass screens are used with light-sensitive film.

4. Transfer technique: metal foils. These are foils which are placed in the neutron beam, become radioactive in proportion to the intensities in the spatial neutron image, and are then removed from the neutron beam and placed in contact with a photographic film. The radioactive emissions from the screen then produce an image on the film. Indium and dysprosium are the most commonly used materials. A material must be chosen which is rapidly activated and has a rapid decay, so that it can be used again. Indium has a half-life of 54 min and dysprosium of 140 min.

5. Track-etch detectors. These detect charged particles by the radiation damage caused in dielectric materials such as polycarbonate plastic sheet, or cellulose nitrate sheet, using an α-emitting converter screen. The radiation damage is made visible by etching in hot sodium hydroxide solution. The image is difficult to view directly, but can be printed with an offset light source. Track-etch screens require high doses but have very good image resolution.

6. Television systems (electronic imaging). Group 1 and 2 radioscopic equipment (section 11.3) can be used with a neutron-sensitive converter screen such as gadolinium oxysulphide or lithium-6 activated zinc sulphide to produce a neutron radiographic image on a TV monitor screen with a much smaller neutron dose than is needed for film radiography [2, 3]. Such a screen image is likely to be grainy due to the polycrystalline nature of the converter screen and will have some gamma-contamination. Some of the loss of image quality can be compensated by frame integration. Such a system can be used for dynamic studies.

The advantage of the transfer technique is that the photographic material does not go into the neutron beam, and as this beam invariably contains a proportion of gamma-radiation, the film is not affected by this gamma-radiation and a purely neutron image is obtained. With a direct-exposure film/screen technique great care has to be taken to minimize the gamma-radiation component of the radiation reaching the film.

Sources of neutrons

There are several potential sources of intense neutron beams, nearly all of which produce fast neutrons which must be moderated to produce thermal neutrons. Possible sources of neutrons are:

1. atomic reactors;
2. positive-ion accelerators (deuteron bombardment of beryllium, tritium);
3. high energy X-ray machines using (X, n) conversion, e.g. linacs;
4. radioactive sources using (α, n) or (γ, n), reactions;
5. radioisotope direct neutron emitters;
6. cyclotrons, synchrotrons.

Occasionally the literature reports the development of special machines with a very high neutron output. After a long development period the superconducting cyclotron outlined by Wilson [4] and Hawkesworth [5] is now in operation for industrial applications [6].

In France and the USA considerable effort has been put into the development of transportable sources. While the 'baby' cyclotron may weigh 3–5 tonnes, a 50 mg californium-252 source or sealed-tube equipment can be built weighing only 500–700 kg, but to date nearly all neutron radiography has been done with neutrons from atomic reactors because they have the ability to provide an intense neutron beam. Californium-252 is a transuranic element which is made artificially. It has a half-life of 2.65 years and a 1 mg source produces 2.3×10^9 neutrons per second.

In positive-ion accelerators, the reaction depends on the accelerating voltage; the 400 kV Van de Graaff equipment uses the $H^3. (d, n). H^4$ reaction. In sealed neutron tubes a target of erbium containing an absorbed deuterium–tritium gas mixture is bombarded with deuterium and tritium ions. A 2.5 MeV Van de Graaff equipment has been marketed in which deuterons are accelerated on to a beryllium target to produce neutrons. Sealed-tube neutron generators, utilizing the $T(d, n)$ reaction, produce 14 MeV neutrons and outputs of up to 10^{12} neutrons per second are possible. The tubes have a limited lifetime of a few hundred hours. Again, these are not in widespread use.

Unless a thermal neutron beam can be extracted from an appropriate part of an atomic reactor, all other sources of neutrons produce fast neutrons which must be moderated and collimated to a useful beam width. This moderation and collimation process entails enormous loss of

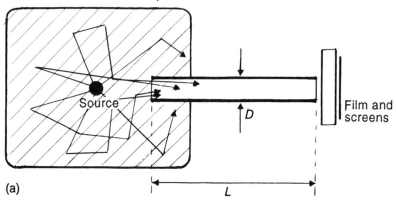

(a)

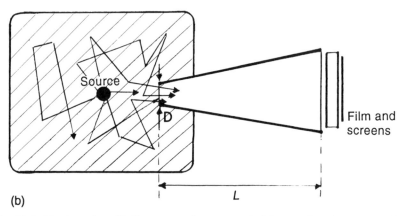

(b)

Fig. 12.1 Neutron 'head'. The source is surrounded by moderating material in which neutrons are slowed by multiple collisions, (a) = parallel-sided collimator; (b) = divergent collimator.

neutrons and the intensity of the useful thermal neutron beam is typically 10^{-5}–10^{-6} of the initial intensity, depending on the L/D ratio (see Fig. 12.1).

To moderate the beam the source of fast neutrons is surrounded by a large volume of moderating material, for example, paraffin wax, oil, water or polythene, in which the neutrons are slowed by multiple collisions. In the moderator volume, therefore, there are thermal neutrons moving randomly, and if a collimator tube containing no moderator is inserted into this volume it will act as an escape route for neutrons, and so become the source (Fig. 12.1). The collimator tube does not necessarily have to point directly at the source of fast neutrons, as shown; its direction and distance into the moderator to give the most intense flux at the location of the specimen and film must be determined by experiment or

Table 12.2 Characteristics of neutron radiographic facilities [7–9]

Source	Basic fast neutron flux (neutrons s^{-1})	Thermal neutron beam	
		L/D ratio	Intensity at film site (no specimen) (neutrons cm^{-2} s^{-1})
Large research reactor	10^{14}	250	10^8
Research reactor	2×10^{11}	100	10^6 cold neutrons
Small reactor	10^{12}	250	10^6
1 mg Cf-252	2×10^9	12	10^4
5.5 MeV linac; Be(X, n)	4×10^8	18	8×10^4
Neutron tube T(d, n)	1×10^8	18	2×10^4
3 MeV van de Graaff	3×10^9	33	2×10^5
15 MeV linac; Pb(X, n)	3×10^{11}	8	1×10^5
Sb–Be, 270 Ci	1.5×10^9	10.5	2×10^4
10 mg Cf-252	2×10^{10}	20	3.5×10^3 cold neutrons

calculation. So far as the film is concerned the effective source of neutrons has a diameter very close to D, at a distance L. Obviously, if L/D is
increased the image definition will be improved (Fig. 12.1(B)) this being
exactly analogous to the (focus size)/(source-to-film distance) calculation in X-radiography. Table 12.2 gives a guide to the typical basic
characteristics of a selection of neutron radiographic sources [7–9].

Cold neutrons are produced by filtering the low-energy end of the
thermal neutron spectrum, or by the use of a spinning 'chopper', or by
using moderator cooling with liquid nitrogen or cryogenic cooling.

In the USA several reactors have been designed specifically to produce
thermal neutron beams for radiography, and because large L/D ratios
are possible, the problem of shielding the film from gamma-radiation is
less severe.

The use of high energy X-ray linacs to produce neutrons, through an
(X, n) conversion in beryllium [10], uranium [11] or lead [8], has been
attempted in only a few laboratories. It has the obvious advantage that
the linac is available for X-radiography most of the time and is converted
to neutron production only when required; this conversion can be quite
rapid, i.e. 1 h or less [10].

Several workers have investigated isotopic neutron sources [7, 12, 13]
and the general conclusion is that californium-252 is the most important
source. Considerable effort has gone into the design of transportable
'neutron cameras' and small-strength sources may be feasible if electronic imaging is used [14].

Cluzeau and others [15–18] reported a mobile radiographic unit,
Diane, which uses a sealed-tube neutron generator GENIE-46, which has

an output of 4.10 neutrons s^{-1} in a 4π solid angle. When moderated in high-density polyethylene (HDP) or beryllium, the thermal neutron output is $1 - 3 \times 10^4$ neutron cm^{-2} s^{-1} 'in the moderator'.

In a rare reference to the use of fast neutrons, Yoshi [19] reported a TV imaging system for fast neutrons (> 2.5 MeV) for the radiography of thick specimens up to 100 mm steel. The neutrons are generated in a small cyclotron and a luminescent converter of polypropylene resin and ZnS is used. The neutrons are the result of conversion of protons in beryllium. The images are said to be 'not as good as with a track-etch detector', but an image can be obtained in 20 s.

In the USA the use of a Cf-252 source with a flux multiplier has been investigated [20, 21]. The primary source is placed near the centre of an array of enriched uranium fuel rods in a water moderator and surrounded by a neutron reflector and a shield. A thermal neutron intensity multiplication factor of $\times 30$ has been achieved and the design is claimed to be fail-safe, but such systems have not yet been licensed for use in Europe.

It has been estimated [6] that 1×10^5 neutrons cm^{-2} on a detector are needed to produce a reasonable image, and 1×10^9 neutrons cm^{-2} for a high quality image, so that Table 12.2 shows very clearly the limitations of all but the atomic reactor sources in terms of neutron flux and exposure times.

In most practical neutron radiography set-ups, except those using neutrons from the largest atomic reactors, a compromise must be made between increasing the L/D ratio to improve image definition (which effectively means increasing D), and keeping exposure times within practical limits. On most non-reactor sources it is essential to use a small L/D ratio so as to have a practical exposure time. With a parallel collimator (Fig. 12.1(A)), D is also the diameter of the maximum area of specimen which can be exposed and must obviously be several centimetres wide to be useful. It is, now common practice, however, to use a divergent collimator (Fig. 12.1(b)) to obtain a larger imaging area from the same effective source diameter, although this creates new difficulties in producing a neutron beam which is not heavily contaminated with gamma-rays.

The design of moderators and collimators for various possible thermal neutron sources was reviewed by Norton [7]. The precise position of the neutron source in relation to the collimator determines the thermal neutron intensity, the neutron/gamma ratio and the neutron spectrum.

Image quality

The image quality of neutron radiographs can be studied in exactly the same terms as have been used for X-rays in Chapters 6–8. Geometric unsharpness with a neutron source is

$$(\text{Object detail to film distance}) \times \frac{D}{L} \qquad (12.1)$$

and film unsharpness is a combination of film and converter screen effects, with the inherent unsharpness of the converter screen usually predominating. Unless a powerful source of neutrons is available, or long exposures are used, it may be essential to use one of the faster film/converter screen combinations, with consequent large unsharpness values and image granularity. Table 12.3 shows unsharpness values and effective speeds for various neutron detectors, using thermal neutrons.

In all except special high-intensity neutron beams from powerful reactors, the geometric unsharpness is likely to be the predominant effect controlling image quality.

Clearly, therefore, the best radiographs with thermal neutrons will be obtained using a thin gadolinium foil screen (or pair of screens), assuming enough neutrons are available, but large values of U_g may make these screens unnecessary. The dysprosium transfer technique has the advantage that it can be used in a gamma-contaminated beam and will still produce a purely neutron image. The zinc sulphide granular screens allow the shortest exposure times to be used, but produce an unsharp and granular image; they are so sensitive to neutrons that their gamma-ray sensitivity is not normally a serious problem, the gamma rejection being × 20 better than gadolinium foil screens.

Track-etch detectors are normally too slow for all except the most powerful neutron beams.

An alternative method of comparing image quality in neutron radiography is the use of MTF values, as discussed in Chapter 8. The MTF curves for the source can be determined from the basic equation

$$R = \frac{\sin \pi f U_g}{\pi f U_g} \qquad (12.2)$$

Table 12.3_Unsharpness values for neutron detectors

Screen	*Film*	U_S (mm)	*Dose required* (neutrons cm^{-2})
ZnS–Li^6 granular	Salt screen type	0.4–1.0	1×10^5
ZnS–Li^6 glass (1 mm thick)		0.4	2×10^6
Gadolinium foil (0.025 mm)	Non-screen	< 0.1	4×10^6
Dysprosium foil (0.15 mm)	Non-screen transfer	0.15	–
Track etch (cellulose nitrate)	–	< 0.1	4×10^8–2×10^9

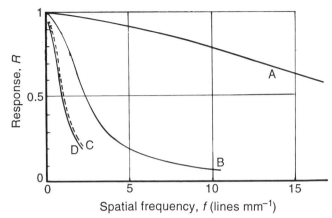

Fig. 12.2 Modulation transfer function curves for various neutron detection systems. Curve A = gadolinium foil direct system; curve B = dysprosium foil transfer system; curve C = ZnS–LiF granular scintillator screen (NE-421); curve D = ZnS–LiF NE905 scintillator glass.

or directly from line spread function measurement on the image of a slit [22]. Similarly, the curves for various types of detector can be determined experimentally, and four curves are shown in Fig. 12.2. If the curves for geometric effects and detector unsharpness are combined, as in Fig. 12.3, one obtains the performance of a neutron imaging system for one particular thickness of specimen and one particular value of L/D. Figure 12.4 shows MTF values for some X-ray energies, and comparison

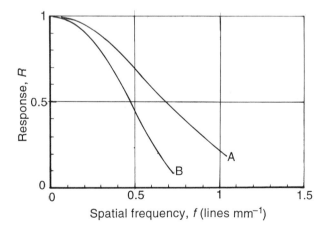

Fig. 12.3 Modulation transfer function curves for neutron radiography using an L/D ratio of 10:1. Curve A = 7 mm thick specimen with NE-905 glass detector; curve B = 13.4 mm thick specimen with NE-905 scintillator glass detector.

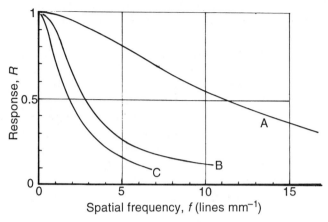

Fig. 12.4 Modulation transfer function curves for X-rays. Curve A = 300 kV X-rays, lead foil intensifying screens; curve B = 200 kV X-rays, calcium tungstate intensifying screens; curve C = 1 MV X-rays with lead screens.

of these with Fig. 12.3 gives a clear indication of the image quality to be expected in neutron radiography with an L/D ratio of 10 and a glass scintillator detector. Similar curves can be built up for other techniques.

With cold neutrons, the images obtained with metal foil screens by the direct method show an outlining effect on some image detail, which enhances its detectability; this effect has not been fully investigated, but it is thought to be due to very low angle neutron scatter.

Applications of neutron radiography

Neutron radiography must be regarded as complementary to X- and gamma-radiography, and unless one has access to a reactor source of neutrons the cost of a suitable neutron source is very high.

One important application is the examination of irradiated reactor fuel material; the radioactivity of the specimen is no great problem if remote handling and image transfer techniques are used, and often suitable neutron beam facilities are available on site.

The second established application is the inspection of small ordnance items for the detection of low-density explosive fillings. The detection of paper separation-washers and plastic components in fuses are all obvious applications for neutron radiography, and there are indications that it is possible to differentiate between different explosive compositions. Inspection of epoxy-bonded metals, to show the absence of filler, is another possible application and there have been claims to detect hydrogen in titanium. There have also been claims for improved detection of corrosion in aircraft, compared with the use of X-rays, and this has been the incentive for the design of portable neutron radiography systems.

There are considerable possibilities in loading material with neutron-absorbent salts to improve detection sensitivity. Thus, soaking a turbine blade in gadolinium nitrate solution, which is absorbed in any casting core residue in the cooling channels, has proved a very successful method of enhancing the image of low-density core material, and there are similar possibilities in other fields. Boron salts are used in some ordnance items, as this is another element with a high neutron attenuation coefficient which will enhance the contrast of the image of the material containing it.

In spite of suggestions that neutrons can be used to penetrate thick metal specimens, there have been few reported applications in this field, probably because with thermal neutrons the penetration is accompanied by considerable scattered gamma-radiation. Neutrons of lower energy than the Bragg cut-off wavelength, however, cannot undergo Bragg scattering in metals so that these materials are virtually transparent to cold neutrons and large penetrations should be possible.

At present, except in the field of radioactive specimens, it seems unlikely that there will be many large-scale applications of neutron radiography which would justify the installation of special equipment when compared with the costs of transport of the stores to a neutron reactor facility. However, there are many smaller applications where a thermal neutron source can provide a useful ancillary inspection technique, so the problem becomes one of economics; is the cost of a small californium-252 source justified by the amount of work?

12.3 PROTON RADIOGRAPHY

If a beam of monoenergetic protons is transmitted through a thickness of material, most of the attenuation occurs after the beam has passed through 90% of the total thickness. It is therefore possible to detect thickness changes of the order of 0.05% [23] by film radiography, which is an order of magnitude better than with X-rays.

West and Sherwood [24] made use of multiple small-angle scattering with protons to enhance the image of an internal edge in the specimen.

It is obviously necessary to have a source of protons, and the proton generator is often a cyclotron; the proton energy must be matched to the specimen thickness. Work on proton radiography is therefore limited to those research establishments with a suitable accelerator.

Film and intensifying screens are used for proton radiography, as for X-radiography.

12.4 ELECTRON RADIOGRAPHY

Most materials emit electrons when exposed to X-rays and as these electrons have a very low penetrating power they can be used for the 'radiography' of very thin materials such as paper. The specimen is put against the film, with a lead foil screen on top, and the electrons from the lead produce the image (Fig. 12.5(a)). There is, necessarily, also a weak X-ray image, but if heavily filtered X-rays (300–400 kV) are used the direct X-ray image is negligible. Paper watermarks, fibre distributions and ink overprints have been studied with this technique, which has been used to authenticate rare postage stamps. Surface emission electron radiography is also possible, by utilizing the different electron emission from different materials (Fig. 12.5(b)). The X-ray beam passes through the photographic film and, again, heavily-filtered moderately hard radiation is used to minimize the direct X-ray exposure to the emulsion. Schnitger and Mundry [25] attempted to relate the image contrast obtainable with different X-ray energies and filtrations to the effective thickness of the specimen using the set-up shown in Fig. 12.5(a). Their results show that for material thicknesses less than 10 mg cm^{-2}, as high an X-ray energy as possible, with a thick copper filter, should be used. Very thin specimens of low density such as paper can also be examined with very low voltage X-rays, typically less than 10 kV (Chapter 7).

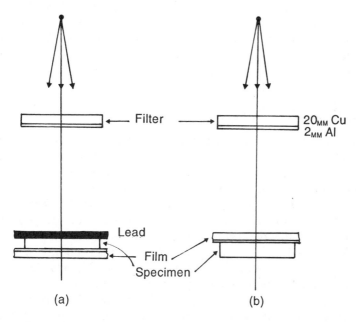

Fig. 12.5 Set-up for electron radiography. (a) = thin specimens; (b) = thicker specimens, surface emission.

12.5 MICRORADIOGRAPHY

This is a technique in which a radiograph of very high resolution of a very thin specimen is obtained, and which is then enlarged photographically to show detail not visible to the unaided eye. The method is, therefore, analogous to optical microscopy. By using X-rays the different absorptions of different parts of a specimen can be utilized to detect, for example, different constituents of an alloy.

Two techniques are used, the contact method and the projection method, the former being the more usual. In the contact method the specimen is placed in contact with the film emulsion and the X-ray tube is some distance away. A very fine-grain emulsion is used, generally 'maximum resolution' plates which can be enlarged photographically 300–500 times before grain begins to limit the resolution. The specimen is reduced to 0.1 mm thickness or less so that very low energy X-radiation can be used, and an X-ray energy is chosen such that the constituents of the specimen lie on opposite sides of one of the characteristic absorption edges of the radiation.

Much of the success of microradiography depends on the careful preparation of the specimen, the uniformity of thickness of the sample and the elimination of scratches. The second important parameter is the choice of the correct X-ray energy to obtain good image contrast, and on metal specimens the most successful work appears to have been done using 15–20 kV X-rays. As the X-ray beam used is not monoenergetic most workers try a range of energies on each specimen. Resolutions of the order of 0.5–1.0 μm are attainable by both contact and projection techniques.

12.6 AUTORADIOGRAPHY

Autoradiography is a method of photographically recording the amount and location of traces of radioactive material within a specimen, this radioactivity being produced either by introducing radioactive constituents or by activating the whole specimen by neutron bombardment.

The specimen is placed in contact with a photographic film, left for a period of time, and the film is then processed. The method can be applied on both macro- and micro-scales, the latter technique requiring special high resolution nuclear-particle emulsions capable of giving the necessary resolution. Macro-autoradiography is relatively simple. It requires good contact between the specimen and film emulsion and there are techniques where a stripping emulsion is coated directly on to the specimen surface, with a thin support. The emulsion is processed while still superimposed on the specimen. 'An example of a typical application is the introduction of ^{32}P into liquid steel during melting; slices of the solidified ingot are then subjected to various heat treatments. Autoradiographs of the surfaces of these slices will show the progress of homogenization with the time of heat treatment.

12.7 RADIOMETRIC METHODS

The use of a radiation detector (Geiger–Müller counter, scintillator, semiconductor) scanned across the specimen and coupled to a recording device, was briefly discussed in Chapter 5. Such systems have been built for specific applications and in some equipment very advanced electronics are used. The basic design points are as follows.

1. The radiation beam should be collimated to the size of the detector; it is good technique to irradiate as small an area of the specimen as practicable.
2. If the image detail is to be improved the size of the scanning spot, i.e. the collimator aperture, must be reduced. If this is done less radiation per unit time reaches the detector, and to get a given quantity of radiation dose at the detector, either the integrating time must be increased or the source intensity made greater.
3. If the radiation dose at the detector is reduced, the signal-to-noise ratio is also reduced, i.e. the noise is greater. Statistical fluctuations always set the limit to the sensitivity attainable.
4. A radiation source of the greatest possible strength should always be used.
5. There is a basic interplay between radiation intensity, speed of inspection and scanning spot size, which must be resolved for each particular application.

In practical circuits it is desirable to compensate for slow variations in intensity and for smooth thickness changes in the specimen, and this can be done by splitting the signal from the detector, putting part through an integrator with a large time-constant, and the other half through an integrator with a short time-constant; the signals are then subtracted and amplified. Automatic gain control circuits are also used.

Nickerson [26] showed by differentiation of the basic absorption equation, that the best sensitivity is attained when $\mu x = 1$, but that the signal-to-noise ratio will be optimum at $\mu x = 2$. In general, radiometric methods have not been widely used because imaging techniques have been preferred.

REFERENCES

[1] Thewlis, J. J. (1956) *Brit. J. Appl. Phys*, **7**, 345.
[2] Berger, H. and Bracher, D. A. (1976) *Proc. 8th World Conf. NDT*, Cannes, France, Paper 3L6.

[3] Lange, H. I., Leeflang, H. P. *et al.* (1994) *Proc. 6th Euro., Conf. NDT*, Nice, France. COFREND, Paris, Vol. 2, p. 817.

[4] Wilson, M. N. and Finlan, M. N. (1986) *Proc. 2nd World Conf. Neutron Radiography*, Paris. Reidel.

[5] Hawkesworth, M. R. (1986) *Proc. 2nd World Conf. Neutron Radiography*, Paris. Reidel.

[6] Boulding, N. J. *et al.* (1994) *Proc. 6th Euro. Conf. NDT*, Nice, France. COFREND, Paris, Vol. 2, p. 795.

[7] Norton, R. J. (1980) Brunel Univ. M.Sc. Dissertation.

[8] Hiraoka, E. and Fujishiro, M. (1978) *Ann. Rep. Osaka Radiation Center*, **19**, 45.

[9] Science Applications Inc., La Jolla, (1977) Commercial literature.

[10] Halmshaw, R. and Hunt, C. A. (1970) *Proc. 6th Int. Conf. NDT*, Hannover, Paper M6.

[11] Gozani, T. (1974) *Trans. Am. Nucl. Soc.*, **19**, 63.

[12] Hawkesworth, M. (1978) AERE Harwell, UK *Rep. R-9211.*

[13] Berger, H. (1970) *Research Techniques in NDT*, Academic Press, London, Vol. 1, Chapter 9..

[14] Meier, H. and Albrecht, W. (1976) *Proc. Conf. Sci. & Ind. Appl. of Cf-252*, Paris.

[15] Cluzeau, S. and Huet, J. (1993) *Nucl. Inst. & Methods in Phys. Res.*, **B79**, 851.

[16] Huriet, J. R. (1986) *Proc. 2nd World Conf. Neutron Radiography*, Paris. Reidel, Vol. 2, p. 1049.

[17] Dance, W. E. (1986) *Proc. 2th World Conf. Neutron Radiography*, Paris. Reidel, Vol. 2, p. 1059.

[18] Cluzeau, S. and Le Tourneur, P. (1986) *Proc. 2th World Conf. Neutron Radiography*, Paris. Reidel, Vol. 2, p. 1059.

[19] Yoshi, K. *et al.* (1993) *J. Nucl. Sci. & Tech.* **30** (12), 1275.

[20] Aerojet Nuclear Corp. (1972) *Californium Prog.* **12**, 52.

[21] Preskitt, C. A. (1972) *Proc. Conf. Sci & Ind. Appl. of Cf-252*, Paris.

[22] Halmshaw, R. (1977) *Brit. J. NDT*, **19** (5), 230.

[23] Koehler, A. M. and Berger, H. (1973) *Research Techniques in NDT*, Academic Press, London, Vol. 2, Chapter 1.

[24] West, D. and Sherwood, A. C. (1972) *Nature*, **239**, 157.

[25] Schnitger, D. and Mundry, E. (1981) *Materialprüfung*, **23** (1), 10.

[26] Nickerson, R. A. (1958) *Non-destructive Testing*, **16** (2), 120.

Index